Colour Atlas of the Surface Forms of the Earth

Helmut Blume
University of Tübingen

Translated by Björn Wygrala

Edited by Andrew Goudie
University of Oxford
&
Rita Gardner
King's College, London

Harvard University Press
Cambridge, Massachusetts 1992

Printed in Great Britain

10 9 8 7 6 5 4 3 2 1

Library of Congress Cataloging-in-Publication Data
Blume, Helmut, 1920–
[Relief der Erde, ein Bildatlas. English]
Colour atlas of the surface forms of the earth / Helmut Blume: translated by Björn Wygrala: edited by Andrew Goudie and Rita Gardner.
p. cm.
Translation of: Das Relief der Erde, ein Bildatlas, Stuttgart, Germany: F. Enke, c 1991.
Simultaneously published in Great Britain: Colour atlas of the surface forms of the Earth. London: Belhaven Press, 1992.
Includes bibliographical references.
ISBN 0-674-14306-X
1. Landforms—Maps. 2. Geomorphology. I. Goudie, Andrew. II. Gardner, Rita. III. Title. IV. Title: Colour atlas of the surface forms of the earth.
G1046.C2B5 1992 <G&M>
551.4′1′0223—dc20 91-31373
CIP

Author's introduction

Geomorphology, a branch of physical geography, is concerned with the surface forms of the earth. Geomorphology investigates the shape, relationships and formation of the surface relief of the earth. Initially description was the prime consideration, but even during the early phases of the development of geomorphology, morphogenesis, i.e. the actual development of the forms, was investigated. The present emphasis of geomorphological studies is on the investigation of the processes which create surface forms. Research is directed by considerations of landform ecology. It attempts to determine the factors which control the development of the surface forms of the earth and to define their relative importance: factors such as the pre-existing relief, geology, climate, water, soil types, vegetation and anthropogenic effects.

Public interest in geomorphological subjects has increased considerably as both global tourism and global news coverage with a strong visual component provide images of landscapes even in distant, previously inaccessible countries. In addition, there is an increasing awareness that surface forms are an important component of the complex ecosystems of the earth, and that human society - especially in unstable or fragile ecosystems - can, for example, initiate intensive erosional processes resulting in fundamental changes of the surface forms of the earth.

This atlas of surface forms of the earth is intended to provide knowledge of the multiplicity of the surficial forms of the earth. The forms, their creative processes, and the rules which govern their differentiation are described in the text and exemplified with photographs, maps and profiles from all over the world. A complete documentation of all types of geomorphological forms was not attempted; photographs and text are limited to the most important processes and forms.

This book will be of value not only to the layman who is interested in the surface forms of the earth and their formative processes, but also to students of geoscientific subjects as an introduction to geomorphology. References are provided for readers who are interested in more in-depth studies.

Tübingen, 1991 Helmut Blume

Editors' introduction

Language is still a barrier to communication, not least in science. As a consequence various intellectual traditions, cultures and concepts have grown up in various countries. In geomorphology, gulfs have developed amongst French, German and English traditions of the discipline. Diversity is desirable, but ignorance and isolation are not. We are therefore pleased that it has been possible to produce an English language translation of a work that in many ways encapsulates, in an accessible and coherent manner, the German traditions in geomorphology. We believe that this approach, with its emphasis on climate in the genesis of landforms and thus on generations of relief resulting from past climate changes, has much from which geomorphologists brought up in the English-speaking traditions can learn. Likewise, its concern with structural controls on geomorphology is an area that needs to receive greater recognition by British and North American scientists.

Scientific traditions apart this book provides, by means of high-quality photographs, an impressive range of field examples of landforms from many different environments around the world. The scale and beauty of many of the features is well portrayed. We have refrained from changing the meaning of the original text, and believe the translation to be a faithful one. There are some areas where our perspectives would differ, but we have not modified the text to reflect such differences. We have tried to make sure that the appropriate English language terms have been substituted for German ones, except in those cases (e.g. inselbergs) where the German terms have international currency, or where there is no appropriate single term translation. We have also provided an English language guide to further reading. This we hope will be helpful to those readers who would like to learn more about the many features and processes that Professor Blume introduces in this volume.

Andrew Goudie
Rita Gardner
Oxford and London,
August 1991

Contents

1 General classification

The large-scale morphotectonic units of the earth

The total area of the earth's surface is 510 million km², of which only 29 per cent is subaerial, i.e. above the present sea level. The remaining 71 per cent is covered by water.

The hypsometric (elevation) diagram of the earth (Fig. 1) shows that a clear subdivision of the earth's surface according to elevation levels is possible. The earth can be subdivided into six large-scale morphotectonic units which are defined by specific elevation ranges: mountain chains (or fold belts), continental platforms, continental slopes, abyssal plains, oceanic ridges and ocean trenches. These large-scale units, and their distribution and elevation ranges, are an expression of tectonic processes and the associated geological structures of the earth's crust which are currently explained by plate tectonic concepts and models.

The crust is the relatively inflexible shell of the earth, and together with the upper part of the mantle, it forms the so-called lithosphere with a total thickness of 70 to 100 km (45 to 60 miles). The lithosphere is subdivided into plates of various sizes which consist of a crustal and lithospheric component of the upper mantle. Continental crust, which underlies the continents and their shelf areas, has an average thickness of 30 to 40 km (20 to 25 miles), and can be up to 80 km (50 miles) thick below the mountain chains. It consists of acidic (i.e. SiO_2 or silica rich), relatively light granitic rocks. Oceanic crust, which underlies the sea-floors, is only 5 to 8 km (3 to 5 miles) thick, and is composed of basic (i.e. with a relatively low silica content) lava rocks with higher densities. Most plates do not consist entirely of either continental or oceanic crust, but rather of a mixture of both types.

The plates are not static; their movement is driven by convection cells in the lithosphere. Upwardly-moving convection streams bring ultrabasic melt from the upper mantle to the surface. These upwardly-moving parts of the convection cells are located below mid-oceanic ridges (Fig. 2), globally extensive graben zones (oceanic rift zones) with elevated margins. In these areas, new oceanic crust is formed by the continual influx of mantle material. As the newly-formed oceanic crust moves away from the ridge on both flanks, so-called 'sea-floor spreading' occurs. The mid-oceanic ridges therefore mark the edges of divergent plates. Initial stages of plate divergence caused by lithospheric perturbations can be seen in continental

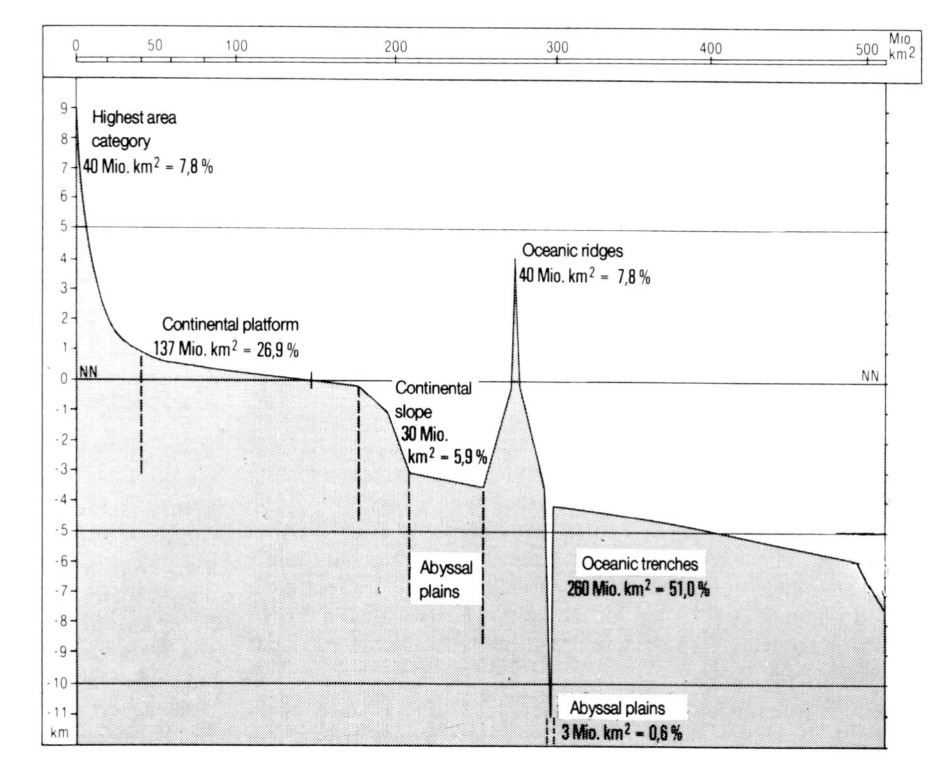

Fig. 1 Hypsometric (elevation) diagram of the surface of the earth (after Leser & Panzer 1981 and Louis & Fischer 1979)

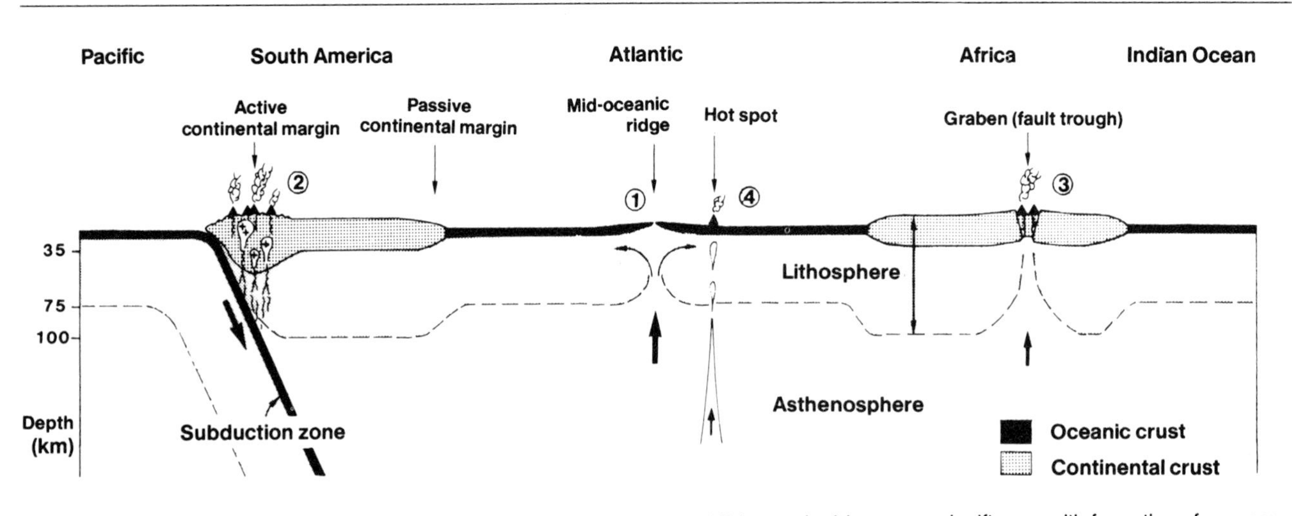

Fig. 2 Plate tectonic units of the earth's crust illustrated by a schematic section through the lithosphere of the southern hemisphere from the Pacific to the Indian Ocean (from Frisch & Loeschke 1986)

The lithospheric plates, which consist of the crust and the lithospheric zone of the upper mantle, are relatively rigid compared to the plasticity of the asthenosphere. Oceanic lithosphere is created at the mid-oceanic ridges and destroyed in the subduction zones.

Owing to the plate movements, active volcanism occurs in four zones of weakness in the earth's crust, and results in different magma compositions:

1) Mid-oceanic ridges: oceanic rift zone with formation of new oceanic crust. Mostly effusive production of tholeiit-basaltic melts with 10–30 per cent upper mantle content.
2) Active continental margins and island arcs: subduction zone volcanism in converging plate areas. Explosive/effusive mixed production of basic and frequently intermediate melts.
3) Continental graben: continental rift volcanism. Explosive/effusive mixed production of basic to acidic, mostly intermediate melts.
4) Hot spots: oceanic intraplate volcanism. Mostly effusive production of tholeiit-basaltic and alkali-basaltic melts.

rift systems, for example the East African Rift Valley or the Red Sea.

Plate divergence initiated in the mid-oceanic ridges and the spreading of the oceanic crust beneath the sea-floors leads to plate collisions at converging plate boundaries. As a consequence, plates are drawn down below other plates in convection streams in so-called subduction zones. This process is dramatically illustrated in the deep ocean trenches. Subducted oceanic crust can be transported to great depths within the earth's mantle due to its high density. The subduction of oceanic crust below continental crust at converging plate boundaries is a prerequisite for mountain-forming processes (orogeny). Mountain chains are formed above the subduction zones along the continental crust margins. The collisions (crustal compression, see Chapter 3) generate fold and thrust structures (Andean type), crustal thickening occurs and new continental crust is created. The boundaries between oceanic and continental plates, where oceanic plates are subducted below continental plates, as for example on the margins of the Pacific Ocean, are defined as active continental margins. They are characterized by abundant seismic and volcanic activity (volcanism of subduction zones, see Chapter 4). Deep-sea trenches and associated island arc systems are typical features of subduction-induced orogeny in the Western Pacific (island arc type). When continental plates collide, as for example in Eurasia, mountain chains are formed, which are similar to the Andean type of orogeny but are not associated with deep-sea trench systems (Alpine type). The Alpine type of plate convergence is also characterized by seismic and volcanic activity.

In addition to diverging and converging plate boundaries, so-called conservative plate boundaries exist at which neither subduction occurs nor is new continental crust formed. In these cases the plates slide past one another along so-called transform faults, as for example at the San Andreas fault zone in North America. These areas are marked by high levels of seismic activity.

The boundaries between oceanic and continental plates do not always coincide with continental margins. As already mentioned, plates can be composed of oceanic and of continental crustal portions which are firmly joined together. The boundary zones between continental crust and oceanic crust within a plate are called passive continental margins and coincide with the previously described morphotectonic unit of the continental slopes. They form the margin of the Indian and Atlantic Oceans with the exception of the Caribbean Sea. Continental shelf areas are of considerable extent when associated with passive continental margins.

The most important features of the morphotectonic units can be summarized from a plate tectonic point of view as follows (Figs. 1 and 2).

Mountain chains

Mountain chains, which consist predominantly of continental crust, are included in the highest area category as defined by an elevation in excess of 1000 m. They can reach altitudes of more than 8000 m (maximum 8847 m, Mount Everest, Himalayas) and represent areas of extensive and comparatively recent uplift with fold and thrust structures at converging plate margins (see Chapter 3). They trend predominantly from West to East in Eurasia and from North to South in North and South America and only comprise a relatively small percentage of the earth's surface. The highest area category covers a total surface area of 40 million km², i.e. 7 · 8 per cent of the earth's surface, with 4 · 7 per cent at altitudes between 1000 and 2000 m and 3 · 1 per cent over 2000 m.

Continental platforms

The morphotectonic unit with the second largest total surface area comprises areas with altitudes ranging from −200 to 1000 m. It includes the shelf zone, i.e. the shallow marine zone, to a depth of 200 m. The continental platforms include several different geological structural types (see Chapter 3), and mostly consist of continental crust. Continental platforms with an area of 137 million km² therefore cover 26 · 9 per cent of the earth's crust.

Continental slopes

This morphotectonic unit, which includes deeply-cut submarine canyons, ranges between the depth levels of −200 m and −3000 m. Continental slopes represent the transition zone between continental and oceanic crust within a plate. They are passive continental margins as previously defined and cover a total surface area of 30 million km², i.e. 5 · 9 per cent of the earth's crust.

Deep sea areas

Deep sea areas (abyssal plains) were created by sea-floor spreading from the mid-oceanic ridges and are totally underlain by oceanic crust. They are subdivided into separate basins and occupy the depth zone between −3000 and −7000 m and, with a total surface area of 260 million km², cover 51 per cent of the earth's crust. They surround the morphotectonic units of the mid-oceanic ridge systems and the deep-sea trenches.

Oceanic ridges and island arcs

Basic mantle material reaches the crustal surface in the mid-oceanic ridges, leading to the generation of new oceanic crust. This occurs at diverging plate margins. A relatively small percentage of the mid-oceanic ridges, for example the island of Iceland, breach the sea surface. Others of the volcanic oceanic islands, which frequently form chains or island arcs, are located on submarine highs. These islands have been formed above localized bulges of the upper mantle, so-called hot spots, to which they are connected by vertical feeder channels. As the lithospheric plates drift slowly across the hot spots, magma flow from the mantle is periodically interrupted and the volcanic island becomes quiescent. Later, however, a new volcano emerges over the hot spot in a new location on the moving plate. On Hawaii the volcanic structures of the oceanic islands rise to a height of 4000 m above sea level (see Chapter 4). After extinction they are subjected to subaerial erosion, often followed by submarine erosion, and can be submerged at great depths where they remain as sea-mounts or guyots. The mid-oceanic ridges and oceanic islands, with a combined surface area of 40 million km², cover 7 · 8 per cent of the earth's crust. The elevation range of this morphotectonic unit is considerable − from −3500 to +4200 m.

Oceanic trenches

Oceanic trenches mark convergent plate boundaries (active continental margins) with active subduction. As previously mentioned, they are areas of strong seismic and volcanic activity as well as orogenic processes. Their depth ranges from −4000 to −1 000 m (maximum −11 033 m in the Challenger Deep south of the Mariana Islands). Their total area is 3 million km², which equals 0 · 6 per cent of the earth's crust.

Form and structure of the large-scale morphotectonic structures of the earth are endogenic, i.e. they are created by internal forces. As soon as they emerge above sea-level they are exposed to the atmosphere and therefore to exogenic (external) forces. These mainly affect two of the morphotectonic units, the mountain chains and the continental platforms, with the exception of the shelf areas and a relatively small part of the mid-oceanic ridges and oceanic islands. The processes of weathering, erosion and deposition, which are predominantly climate-controlled, then determine the multitude of different surface forms in the subaerial regime of the earth's surface. They are described in Chapter 2. The effects of these exogenic forces on the endogenically-created landforms are not only felt after the cessation of the endogenic activities, but already occur contemporaneously with the tectonic processes during long geologic time spans. The exogenic forces do not simply sculpture the completed endogenic surface forms, but attack the endogenically developing forms as soon as they emerge above sea-level. It should be added that the submarine morphotectonic units are also subjected to erosion and accumulation processes.

Zoning and vertical differentiation of landforms

The exogenic or geomorphological processes, and the specific landforms which they create, are zoned both latitudinally and atlitudinally. These zones are a function of the climate distribution patterns on the earth's surface. As a broad generalisation, suggested by Köppen, the earth can be subdivided into a sequence of climatic zones between the poles and the equator which are mainly defined by their thermal regime, i.e. by their temperatures: polar and subpolar zones, cold and cool temperate climatic zones, warm temperate subtropical zones, and finally the tropical zone. These basically thermal climatic zones can then be subdivided according to humidity or rainfall values into arid (dry) and humid (wet) subzones. Frost climate, which in the polar and subpolar zones can predominate at sea-level, is, however, not restricted to these climatic zones. Due to decreasing temperatures at higher altitudes, frost climate zones can also occur at high altitudes in warmer climatic zones including the subtropics and tropics. This general subdivision of the latitudinal and altitudinal climatic zones of the earth's surface is also reflected in the respective differentiations of the geomorphological processes, not only according to erosional types but also to erosional intensities (see Chapters 2 and 5).

Central Europe for example, which is presently characterized by humid, cool-temperate conditions and a generally slight to moderate relief, is only exposed to moderate geomorphological forces. Corresponding relief in other climatic zones can be formed by much stronger geomorphological forces, for example in the marginal tropics with strong seasonal humidity variations, or even by weaker forces as in extremely arid zones.

Owing to the apparently slight erosional processes in the lowland and highland relief of the humid, cool-temperate climate zone, erosional quiescence is often assumed for these areas. But the overall erosional amounts, i.e. the progressive lowering of the entire surface, average 3 cm per 1000 years. These denudation rates can be determined by measuring the amount of material which is transported in a specific drainage area. They range from 1 to 4 cm per 1000 years in lowland and highland areas, but can reach 20 to more than 70 cm per 1000 years in high-altitude and high-relief areas within the same climatic zone. The extent and intensity of the geomorphological processes is readily apparent. Consequently, erosion in the high mountain regions is a much more important process than in the lower altitude areas with their slight to moderate relief. This applies to all of the earth's climatic zones.

Moreover, the geomorphological processes differ between these lowland and highland areas. Table 1 shows clearly that in low to moderate relief regions, transport in solution is the dominant mode, whereas in high mountain areas, detrital fluvial transport plays the most important role. In humid, cool-temperate climatic zones, eroded material is predominantly moved in solution, a process which is particularly important in areas with easily soluble lithological types (see Chapter 7). Previous assumptions of erosional quiescence in low to moderate relief environments in humid, cool-temperate conditions are therefore not generally valid.

Table 1 Fluvial material transport in humid areas of cool-temperate climatic zones (from Louis & Fischer 1979)

Fluvial material transport	Areas of low and medium relief (per cent)	Areas of high relief (per cent)
Solution load (chemical material)	75–95	10–45
Suspension load (clay and silt)	< 15	40–70
Sand and pebbles	< 15	10–30

Relative dimensions of relief units

Table 2 summarizes the basic subaerial relief units in descending order of size from megarelief to picorelief (miniature forms). Basal extension, surface area and height are used as criteria in order to assign a relief unit to the respective dimensional range. Mega- and macroreliefs correspond to the structural types which can be recognized within the large-scale morphotectonic units (see Chapter 3). As in the case of the smaller relief units, they are formed by geomorphological processes with both endogenic and exogenic contributions.

Small-scale maps such as those in the covers of this book (scale 1 : 80 million) can only be used to describe mega- and macroreliefs. Large-scale geomorphological maps enable smaller forms to be shown. Fig. 3, for example, shows a section of sheet number 4 (Wehr) of the Geomorphological map of the Federal Republic of Germany (GMK 25) using a scale of 1 : 25 000. This map includes all relief units down to a basal size of 100 m (microrelief). The next larger relief unit, the mesorelief, cannot be read directly from the map. It can, however, be interpreted indirectly from the individual relief elements or areas with common properties. The GMK 25 map sheet contains the following geomorphological information: slope angles, curvature, breaks and steps, valleys and drainage ways, minor landforms, substrata, geomorphological processes and areas, hydrographic data and supplementary information. The GMK 25 map of detailed geomorphological features is classified according to both processes and morphology.

The 1 : 100 000 scale geomorphological map of Germany (GMK 100), a section of which is shown in Fig. 4, was also constructed using unitized construction principles, in which larger relief units were resolved into relief elements. Due to the smaller scale compared to GMK 25, only relief units with a basal width of more than 500 m were resolved into relief elements, and GMK 100 is therefore a more generalized map. Map scales are also the reason for employing summary symbols on GMK 100 instead of showing each

Table 2 Dimensions of mainland relief units (from Barsch & Stablein 1978)

	Size (B) (basal width)		Surface (F)	Height (H)	Examples
Megarelief (very large forms)	above 10^6 m (= 1000 km)		above 10^{12} m^2 (= 1 Million km^2)	—	Canadian Shield
	10^6 m	Limit	10^{12} m^2		
Macrorelief (large forms)	approx. 10^5 m (= 100 km)		approx. 10^{10} m^2 (= 10 000 km^2)	—	Alps Folded Jura Rheingraben Harz
	10^4 m	Limit	10^8 m^2	10^3 m	
Mesorelief (medium forms)	approx. 10^3 m (=1 km)		approx. 10^6 m^2 (=1 km^2)		Kyffhäuser Hills Kieler Förde (see photo 10.14) Rochusberg/Bingen, Germany
	10^2 m	Limit	10^4 m^2	10^1 m	
Microrelief (small forms)	approx. 10^1 m (= 10 m)		approx. 10^2 m^2 (= 100 m^2)		Dead ice hollow Doline Dune
	10 m	Limit	10 m^2	10^{-1} m	
Nanorelief (very small forms)	approx. 10^{-1} m (= 10 cm)		approx. 10^{-2} m^2 (= 100 cm^2)		Karren Tafoni
	10^{-2} m	Limit	10^{-4} m^2		
Picorelief (miniature forms)	below 10^{-2} m (= 10 mm)		below 10^{-4} m^2 (=1 cm^2)	—	Glacial striations

individual form as on GMK 25. GMK 100 obviously shows less detail than the larger-scale GMK 25. The GMK 100 concept is designed to emphasize synthesized data, general types and morphogenetics.

The differences are quite obvious when Figs. 3 and 4 are compared. Fig. 3 (large scale) shows the eastern section of the Dinkelberg, a marginal block in the southern Black Forest. It consists of limestones (*Muschelkalk*) with a karst relief formed by dolines (sinkholes). In the GMK 25 map, every doline is shown with its correct size on the plateau, which is defined as 'denudative' by the appropriate colour; in contrast, the medium-scale GMK 100 map (Fig. 4) features a generalized plateau description (indicated by the 'karst' colour), but the doline symbols give no indication of the actual size of the dolines – they are just symbols. As GMK 100 places more emphasis on morphogenetic characteristics than GMK 25, it also features, for example, the fault scarp to the north of which the basement rocks of the Black Forest rise; this feature is not marked on GMK 25.

Relief generations

Direct measurements made in modern geomorphological studies can only qualify and quantify recent geomorphological processes. As already mentioned, these only play a subordinate role in explaining the landforms present in humid cool-temperate climatic zones as opposed to their importance in other climates. Present landforms, for example in Central Europe, cannot be explained by Recent geomorphological processes, with Recent being defined as the time span since the end of the last Pleistocene glacial period, i.e. approximately 10 000 years before present.

Recent (or present) landforms in humid cool-temperate climatic zones are created by erosional incision of valley floors which were covered by alluvium during the last glacial period, and by weak denudation processes on slopes (see Chapter 2). Surface forms in general in Central Europe must therefore be related to formative processes which were active during different paleo-climatic conditions. The surface forms are preformed (or paleo-forms), and can be subdivided into different relief generations. A relief generation is defined as all landforms which are simultaneously created in an area by local climatic conditions. As climatic conditions have changed in various ways in all areas of the earth, several relief generations can usually be discerned and relief is therefore mostly of polygenetic, or multi-phase origin.

In Central Europe, formative processes on slopes and on valley floors, even if they are only very weak, can be defined as the youngest relief generation. These have only slightly modified landforms inherited from older relief generations. Two main relief generations control landforms in Central Europe. The principal features of the Pleistocene relief generation, which commenced 1 · 8 mybp (million years before present), are the valleys which were episodically excavated during the warmer phases of the last 600 000 years, and the marked modification of the older Tertiary landforms by solifluction processes during glacial periods. The tertiary forms are the oldest relief generation in Central Europe; lower lying areas were probably created as pediments during phases with subtropical conditions which were preceded by tropical conditions. Remnant surfaces at higher elevations were probably formed during periodically changing humid tropical conditions (see Chapter 5). The sequence of relief generations can obviously vary in different climatic zones.

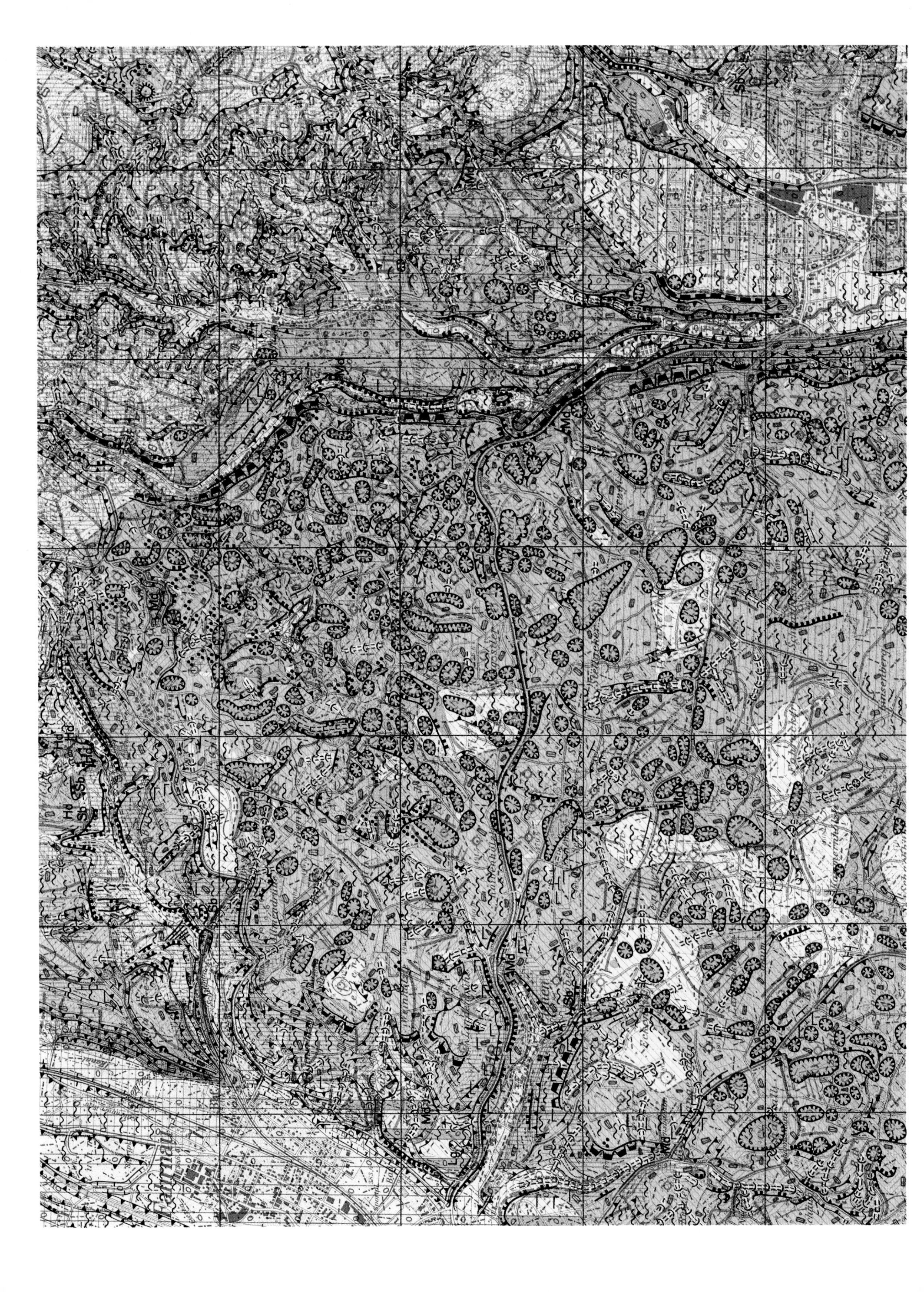

1 Neigungen
slope angle

- 1.1 0°-0,5°
- 1.2 > 0,5°-2°
- 1.3 > 2°-7°
- 1.4 > 7°-11°
- 1.5 > 11°-15°
- 1.6 > 15°-35°

2 Wölbungslinien
axes of curved slope segments

	konvex *convex*	konkav *concave*	Radius des Krümmungskreises *radius of curvature*
2.1			6-< 300 m
2.2			300-600 m

3 Wölbungen von Kuppen und Kesseln
curvature of hillocks and depressions

Radius des Krümmungskreises *radius of curvature*

- 3.1 Vollform < 300 m — *crossing point of convex curvatures*
- 3.2 Hohlform < 300 m — *crossing point of concave curvatures*
- 3.3 Vollform 300-600 m — *crossing point of convex curvatures*

4 Stufen, Kanten und Böschungen
steps and breaks of slope

	Stufenhöhe [m] *heigth of step*	Grundrißbreite [m] *width of step*
4.1	<1	1-5
4.2	>1-5	1-5
4.4	>5-20	1-5
4.5	>5-20	>5-10
4.6	>5-20	>10
4.7	>20	

- 4.8 Schichtstufe (B < 100 m, H > 20 m) — *cuesta*

5 Täler und Tiefenlinien
valleys and small drainage ways

- 5.1 Muldental (Breite 25-< 100 m) — *valley with gently sloping concave sides*
- 5.2 Sohlental (Breite 25-< 100 m) — *box-shaped valley*
- 5.3 Kerbtal (Breite 25-< 100 m) — *V-shaped valley*

6 Kleinformen und Rauheit
minor landforms and roughness

- 6.1 Kuppe — *dome*
- 6.2 Kessel, Doline — *doline*
- 6.4 Nische — *niche*
- 6.8 Fächer, Kegel — *fan*
- 6.10 Lesesteinhaufen — *floats*
- 6.12 wellig — *wavy surface*
- 6.13 höckerig — *hillocky*
- 6.14 kesselig — *kettled*
- 6.15 stufig — *stepped*

8 Substrate
substratum

- 8.8 Lehm, Lößlehm, verlehmte Schotter — *loam, loessloam, loamy gravels*
- 8.9 Schuttdecken mit Kalksteinbraunlehmen — *covers of debris with brown loam of limestone*
- 8.11 sandiger Lehm mit Geröll — *sandy loam with gravels*
- 8.13 sandige Lehmmatrix mit Schutt < 200 mm Kantenlänge — *matrix of sandy loam with debris < 200 mm length of side*

10 Schichtigkeit des Lockermaterials
layering of subsurface materials

- 10.1 Unterlagerung — *underlaying material*
- 10.2 Überlagerung — *overlaying material*

12 Geomorphologische Prozesse
geomorphological processes

- 12.2 flächenhafte Abspülung — *sheet wash erosion*
- 12.14 Arbeitskanten an Fließgewässern — *working edge in stream beds*

13 Geomorphologische Prozeßbereiche
areas of geomorphological processes

- 13.1 äolisch — *aeolian*
- 13.2 karstisch / subrosiv / korrosiv — *karstic / subrosional / corrosive*
- 13.4 cryogen — *cryogenic*
- 13.5 fluvial — *fluvial*
- 13.6 denudativ — *denudational*
- 13.7 anthropogen — *anthropogenic*
- 13.9 chronologische Verschiedenheiten bei fluvialen Sedimenten: hell = jünger, dunkel = älter — *chronological differences of fluvial sediments: light = younger, dark = older*

14 Hydrographie
hydrography

- 14.2 Gewässer, perennierend — *perennial streams and lakes*
- 14.3 Gewässer, zeitweise fließend — *streams and lakes, intermittent flowing*
- 14.5 künstliches Gewässer, ständig fließend — *perennial artifical drainageways*
- 14.11 Staunässe — *impeded drainage*
- 14.13 Quelle, ständig fließend, ungefaßt — *spring, perennial, not catched*
- 14.14 Quelle, ständig fließend, gefaßt — *spring, perennial, catched*
- 14.19 Stromschnelle, Wasserfall — *rapid, waterfall*

15 Ergänzende Angaben
supplementary informations

- 15.3 Lg Lehm- und Tongrube — *loam and clay pits*
- 15.4 Md Mülldeponie — *rubbish dump*
- 15.6 Sb Steinbruch — *quarry*

Fig. 3 Part of the Geomorphological map 1 : 25 000 of the Federal Republic of Germany (GMK 25), Sheet 4 (Wehr), Berlin 1979

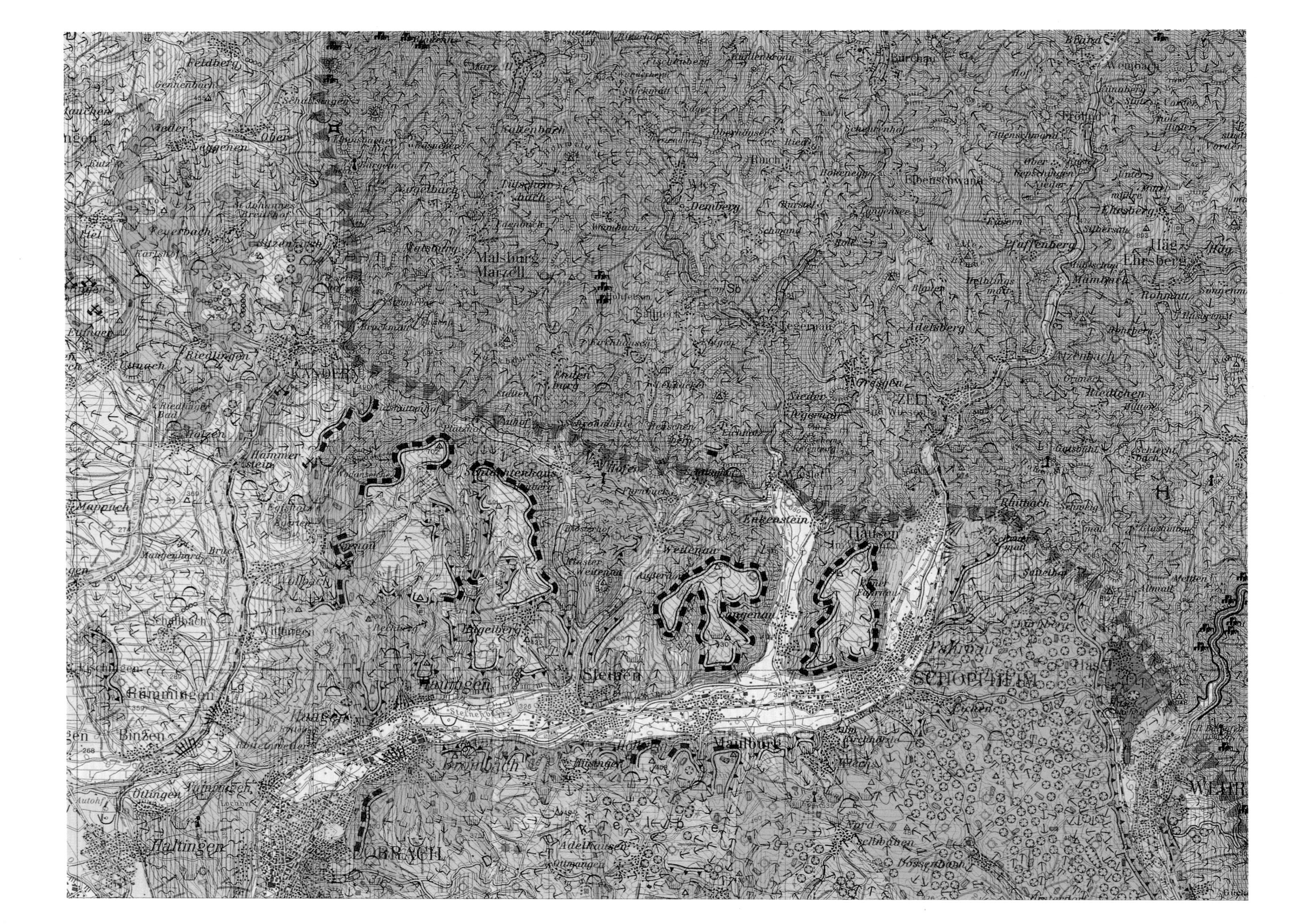
SCHOPFHEIM
LÖRRACH
WEHR
Steinen
Maulburg
Malsburg
Marzell
Feldberg
Binzen
Haltingen
Riedlingen
Zell
Hausen
Enkenstein
Weitenau
Dernberg
Elbenschwand
Pfaffenberg
Adelsberg
Fahrnau
Adelhausen
Brombach
Rümmingen
Wittlingen
Tegernau
Ehsberg

1 Neigungen *slope angles*

- 1.1 < 2°
- 1.2 > 2°-4°
- 1.3 > 4°-7°
- 1.4 > 7°-15°
- 1.5 > 15°-35°

2 Wölbungen *axes of curved slope segments*

konvex *convex* / konkav *concave*

- 2.2 Wölbungsradius 6-300 m *radius of curvature*
- 2.3 Wölbungsradius > 300 m *radius of curvature*

3 Kanten und Geländestufen *steps and breaks of slope*

Höhe *height*

- 3.1 > 1-5 m
- 3.2 > 5-10 m
- 3.3 > 10-20 m
- 3.4 > 20-100 m
- 3.6 Bruchstufe (topographischer Höhenunterschied > 100 m) *fault-line scarp (difference in elevation > 100 m)*
- 3.7 Schichtstufe *cuesta scarp*

4 Talformen *valleys*

Kleine Täler, obere Spannweite bis 500 m *small valleys, upper width up to 500 m*

- 4.2 Kerbprofil (symmetrisch/asymmetrisch) *V-shaped valley (symmetrical/asymmetrical)*
- 4.3 Muldenprofil (symmetrisch/asymmetrisch) *saucer-shaped valley (symmetrical/asymmetrical)*
- 4.4 Sohlenprofil (symmetrisch/asymmetrisch) *flat-floored valley (symmetrical/asymmetrical)*

Täler, 500 m-1000 m obere Spannweite *valleys, upper width from 500-1000 m*

- 4.5 Muldenprofil (symmetrisch/asymmetrisch) *saucer-shaped valley (symmetrical/asymmetrical)*
- 4.6 Sohlenprofil *flat-floored valley*
- 4.7 Täler mit natürlich terrassiertem Talquerschnitt *valleys with terraced profile of natural origin*

6 Kleinstformen (Rauheit), Klein- bis Mittelformenbereiche *micro, minor and mesoforms*

- 6.1 rillig *furrowed*
- 6.6 Schwemmfächer *alluvial fan*
- 6.7 Trichterfeld, Kesselfeld (Dolinen) *pitted area (dolines)*

7 Geomorphographische Einzelkennzeichen *signs of geomorphographical singularities*

- 7.1 Kuppe *knoll*
- 7.3 Sporn *spur*
- 7.6 Damm *earth dam*
- 7.7 Felsköpfe *rock outcrops*
- 7.9 Trichter, Doline *funnel-shaped form, doline*
- 7.11 Sattelpunkt *saddle*
- 7.12 Gipfelpunkt, mit Höhe in m NN *peak, height in m NN*
- 7.13 Wölbungspunkt, konvex, r < 300 m *hillock, convex, r < 300 m*

8 Geomorphologische Prozeß- und Strukturbereiche *geomorphological process and structure areas*

- 8.1 Löß/Lößderivate *loess/loess derivatives*
- 8.2 karstisch *karstic*
- 8.4 kryogen *cryogenic*
- 8.5 fluvial: Auebereich *fluvial: flood-plain*
- 8.6 fluvial: Niederterrassenbereich *fluvial: Low Terrace*
- 8.7 fluvial: Hochterrassenbereich *fluvial: Main Terrace*
- 8.9 denudativ *denudational*
- 8.10 deluvial *deluvial*
- 8.11 fluviale Ausraumzonen, denudativ überprägt *areas of fluvial erosion modified by denudational processes*
- 8.12 strukturell *structural*
- 8.13 Kar oder karähnliche Bereiche *cirque or nivation hollow*

11 Lockersubstrattypen *subsurface material*

- 11.3 Sanddecken mit Feinschutt, geringmächtig *sand cover with fine debris, shallow*
- 11.4 Sandsteinschuttdecken, geringmächtig *sandstone debris cover, shallow*
- 11.8 Schotter- und Kiesdecken, meist sehr mächtig *gravel cover, in general very thick*
- 11.12 Lehm- und Kolluvialdecken aus Löß oder Lößderivaten, verschiedenmächtig *loam and colluvial cover of loess or loess derivatives of different thickness*
- 11.13 Löß oder Lößderivate auf würmzeitlichen Schottern, verschiedenmächtig *loess or loess derivatives on Wurmian gravels, of different thickness*
- 11.15 Kalkschuttdecken, meist sehr flachgründig *limestone debris cover, in general very shallow*
- 11.18 Ton-Mergeldecken mit verschiedenen Schutten, oft tiefgründig *clayey marl cover with varying debris content, often thick*

13 Hydrographische Verhältnisse *hydrography*

- 13.1 Gewässer, perennierend, Fluß *perennial stream, river*
- 13.2 See, Tümpel, perennierend *perennial lake, perennial pond*
- 13.3 künstlicher See *man-made lake*
- 13.5 Gewässer, zeitweise fließend *intermittent stream and pond*
- 13.6 Gewässer, künstlich verändert, zeitweise fließend *intermittent stream, modified by man*
- 13.7 natürliches Gewässer, perennierend, künstlich verändert *natural perennial stream, modified by man*
- 13.8 künstliches Gewässer, ständig fließend *perennial artificial drainageway*
- 13.9 künstliche Gewässer, zeitweise fließend *artificial drainageway*
- 13.12 vernässt *waterlogged*
- 13.13 Quelle, ständig und zeitweise fließend *perennial and intermittent spring*
- 13.17 Wehr, Staustufe *weir*
- 13.19 Wasserbehälter *water tank*
- 13.20 Kläranlage *sewage plant*

14 Anthropogene Einzelformen *single anthropogenic forms*

- 14.1 Hd Halde *dump*
- 14.3 Hohlweg *sunken road*
- 14.5 Kg Kiesgrube *gravel pit*
- 14.7 Mülldeponie *rubbish dump*
- 14.8 Sb Steinbruch *quarry*
- 14.9 Burgwall, Ringwall, Schanze *castle's wall, ring wall, fortification*
- 14.10 Aussichtsturm, Aussichtspunkt *lookout tower, observation point*
- 14.12 Bergwerk, aufgelassen *mine, abandoned*

15 Anthropogen beeinflußte Flächen *anthropogenically influenced areas*

- 15.3 durch Anbau (z.B. Weinbau) umgestaltet *modified by cultivation (e.g. viticulture)*

Fig. 4 Part of the Geomorphological map 1 : 100 000 of the Federal Republic of Germany (GMK 25), Sheet 2 (Freiburg-Sud), Berlin 1985

2 Processes: weathering and erosion

The entire surface of the earth (i.e. all areas above sea-level, see Chapter 1) is exposed to the exogenic processes of weathering and erosion, with the exception of some forms which are created by exogenic depositional processes. The basic forming processes are fluvial, glacial, aeolian and marine, and these are discussed in Chapters 5, 8, 9 and 10 respectively.

Weathering

Weathering is defined as the surficial processes which affect the mineral cohesion of rock materials, for example, by altering, weakening or destroying them. By reducing the strength of the rocks, unconsolidated material (weathered mantle, detritus) is provided which can then be transported by erosional processes (Fig. 5). Weathering processes are influenced by lithological types, but are predominantly climate-controlled. The decisive factor for the resulting morphology and relief of a particular rock type is its petrographic characteristics, such as its chemical and mineral composition, permeability, fracturing, bedding or cleavage. These parameters control the resistance of the rocks.

The following rock types are generally considered to be more resistant to weathering forces:
Magmatic rocks: nearly all basement rock types (e.g. granites and diorites) and most extrusive rocks (e.g. phonolites and basalt).
Sedimentary rocks: sandstones, especially silica-cemented.
Metamorphic rocks: gneiss, quartzite and marble.

Lithology types which are less resistant to weathering include:
Magmatic rocks: volcanic ash and tuff.
Metamorphic rocks: shales and mica schists.
Sedimentary rocks: claystones and marl; limestones with respect to their solubility (see Chapter 7), while their resistance to physical weathering is quite high.

Weathering is the result of both physical and chemical processes.

Physical weathering

Physical weathering is also often described as mechanical weathering, as it leads to mechanical disintegration of the original rock material. The resulting weathered residue has components of various sizes, from blocky, coarse or fine debris, to gravel, sand and dust. Insolation weathering (Photo 2.1), which is also described as direct temperature weathering, is caused by the daily cycle of heat input from solar radiation and loss of heat at night, i.e. constantly alternating heating and cooling phases. Diurnal temperature fluctuations of more than 50°C often occur in tropical and subtropical arid areas. A temperature increase of 50°C leads to thermal expansion for example of 0 · 25 to 0 · 6 per cent in granites and sandstones respectively. This is the equivalent of an increase in surface area of 5–10 cm^2 for a 1m^2 area of rock. Repeated expansion and contraction of the surficial parts of the rocks loosens the cohesion of the mineral constituents and the rock disintegrates. Frost weathering (Photo 2.1) can be defined as indirect temperature weathering. The volume expansion (approximately 9 per cent) during the phase change from water to ice exerts destructive forces in fractured rocks. Diurnal freezing and thawing, which is most frequent at high altitudes and in subpolar regions, is a prerequisite for frost weathering, and fracturing or cleavage will accelerate the process. Rock structures can also be loosened by the pressure exerted by plant roots and by wave action on exposed coastal areas (Photos 10.3 and 10.4). Physical weathering also includes exfoliation processes (Photo 2.3) during which thin flakes or plates are formed and removed from rock surfaces. They are caused by decreasing pressures when rocks are released from their confining subsurface pressure environment, and are common in massive rock types where decompression forms so-called pressure-release jointing.

Chemical weathering

Chemical weathering includes both hydrolytic and solutional processes. This section only covers the hydrolytic destruction of rocks (solution processes are discussed in Chapter 7). Hydrolysis requires the presence of water, and its effectiveness is intensified by increasing temperatures. Hydrolytic weathering (Photo 2.4), which results in considerable elemental exchange processes, refers to the weathering of silicate minerals (and is therefore also called silicate weathering). The silicates, including feldspars and related minerals, comprise 60 per cent of all minerals in the earth's crust. Even though they are relatively insoluble in pure water, silicate minerals are affected by various commonly occurring acids. Preceding mineral hydration (inclusion of water into the chemical composition) increases the effectiveness of hydrolytic weathering. Ferrosiallitic (iron-silica-aluminium rich) or ferrallitic (iron-aluminium rich) residues are produced by the strong hydrolytic elemental exchange processes. The first, non-desilicified type forms a thin mantle in perennially humid tropical margins. The second type can penetrate to great depths in the humid inner tropics and represents the typical final stage of hydrolytic weathering. Hydrolytically formed residues can be recognized by their characteristic red colouring due to abundant iron-oxides in the weathered minerals. Lack of oxygen in the deeper parts is indicated by white or grey colours. In massive rock types (e.g. granites), the unconsolidated, weathered material can contain isolated 'floating' blocks with soft weathered crusts, but which still retain a solid unweathered core. Ferrallitic weathering products, which are hardened into crusts by subaerial exposure, are called laterites.

Physico-chemical weathering

Most weathering processes are a combination of chemical and physical weathering. Hydration is the inclusion of water into the composition and structure of crystals and minerals, and is a common process on rock surfaces and especially along fractures and faults in all climatic zones. Hydration disturbs the crystal lattice and leads to rock disintegration. The process is enhanced by pressures exerted as a result of the volume increase of minerals, and especially of salts, during hydration. The abundant weathering mantles of coarse rock residue in humid cool-temperate climatic zones are largely the result of hydration processes. In warmer climates, hydration precedes hydrolytic weathering.

Repeatedly changing moisture contents (wetting-drying cycles), owing, for example, to dew condensation and evaporation, can cause salt weathering or salt wedging (Photo 2.5). This occurs when water evaporation at the surface, and the concomitant capillary effects, draw saline solutions to or close to the surface. Salt crystallization exerts pressures which lead to the development of thin scaly plates or layers on rock surfaces. A combination of salt and insolation weathering results in the formation and separation of sheets of rock which can be up to several tens of centimetres thick. This process is also described as desquamation (Photo 2.6). Capillary water movements towards the rock surface and the precipitation of dissolved salts near the surface, both caused by evaporation, also form hard crusts. These include calcareous crusts or calcrete (Photo 2.7) and other duricrusts (Photo 2.8). The latter type of crust, as well as desert varnishes, can be composed of manganese, iron or silica.

As opposed to humid climates, where weathering proceeds downwards from the rock surface, arid areas are characterized by chemical processes which are directed from the inside of a weathered rock to its surface. The resultant weakening of the rock core is accompanied by hardening or crustal formation on the surface, but the process does not stop here. As soon as the duricrust is destroyed by physical weathering and the loose weathered material has been removed by wind or water, duricrust formation is immediately renewed. These processes are also related to the formation of caverns or tafoni (as originally described in Corsica; Photo 2.9), which are especially widespread in massive rock types and are mostly associated with duricrust formation. Caverns are formed after the duricrust has been partially destroyed and the originally concealed loose material has been removed. Rapidly alternating moisturizing and desiccation are a prerequisite for this type of weathering.

Weathering caused by seepage water (Photo 2.10), must also be mentioned as it plays a special role in sandstones owing to their higher porosity. Resistance differences between cemented and loosely-bound sandstones, exploited by water seepage, form unusual microforms, for example, honeycomb-shaped or alveolar erosional holes which can cover entire rock faces. A combination of these erosional processes leads to the formation of bowl-shaped weathering hollows or pans (Photo 6.8) without drainage channels, which are found in horizontal or subhorizontal surfaces of massive rock types. These used to be interpreted as prehistoric 'sacrificial bowls' according to popular tradition. Their formation is mainly ascribed to biochemical processes in weak rock zones caused by acidic solutions from organic material such as leaves or moss, together with frost weathering effects.

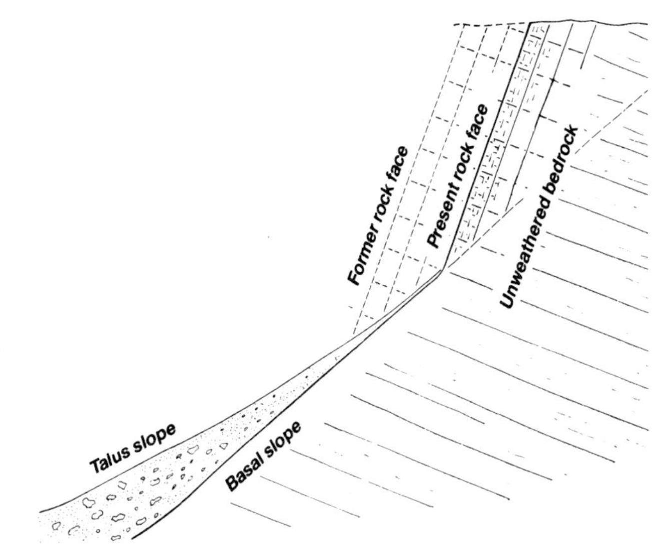

Fig. 5 Slope degradation (from Louis & Fischer 1979)

After sufficiently strong weathering, the rock wall recedes as rock fragments become detached. The base of the wall moves upwards and a basal slope (or wash slope) is created. Detached fragments form a debris mantle which thins towards the base of the rock wall. If the process is not interrupted, the rock wall is completely replaced by a talus slope on the basal slope. If the talus slope material is removed by other processes, the bare bedrock of the basal slope is revealed as the new surface.

Erosion

Many types of erosional processes are involved in the creation of surface forms. Denudation, surficial water flow and linear erosion are mentioned here, while other processes such as solution, glacial, aeolian and marine processes are discussed in Chapters 7–10. Denudation, surficial water flow and linear erosion are the result of a complex matrix of geomorphological processes, in which several factors combine to control the processes to varying degrees. These factors can be subdivided into two main groups. Exogenic factors include climate, hydrology, vegetation, and anthropogenic effects, while endogenic factors are lithologic composition, structure and relief. The last of these, relief, is expressed as both elevation differences within small surface areas and the position relative to the base-level of erosion, and it provides potential energy to the system.

Denudation

Denudation (gravitational erosional processes) can be of a gradual or a spontaneous nature. The gradual, slow processes mostly only result in slight short-term effects. Rockfalls along cliffs, which are subjected to physical weathering, lead to cliff degradation (Fig. 5) with basal debris accumulations (talus slopes; Photo 2.11). Fractured sandstones are particularly prone to the formation of isolated rock pillars (Photo 2.12). Highly resistant rock-beds weather more slowly than less resistant beds and form protruding ridges or even overhanging cliffs (Photo 2.13). Varying degrees of rock resistance can also lead to mushroom-shaped rock forms (Photos 2.14, 9.1), while natural arches (Photo 2.15) are an additional typical relict of sandstone cliffs. If hydration weathering has created a mantle of weathered debris, gravitational mass flow in the form of soil creep occurs on slopes. These movements down slopes can be recognized by typical bulges where flow has been constrained by solid bedrock blocks, or by tree trunks. Soil creep can also be recognized by distorted bedding planes, that is when bedding planes, which originally were vertical to the slope plane, are deflected down towards the base of the slope. In the case of soil movements in periglacial conditions (gelisolifluction; Photo 2.16), soil creep is a relatively powerful erosional agent (see Photo 5.21).

Spontaneous and episodic gravitational erosional processes move large amounts of material within very short time spans. Fall, slide, glide or slip processes can occur. Rock falls (Photo 2.17) occur on steep slopes and especially on rock walls. The largest known rock fall, during which 20km² of rock material was moved, occurred in the Iranian Zagros mountains. Slide processes occur as block slides in massive rocks, as block glides when massive rocks move over weaker rocks (Fig. 6, Photo 2.18), and as slips (Photos 2.19, 2.20) in unconsolidated material. In mountainous terrains, under conditions of high pore water pressures, the high relief energy leads to the release of debris flows (Photo 2.21) or mud flows which can be differentiated from the other spontaneous types of mass transport by their high water content.

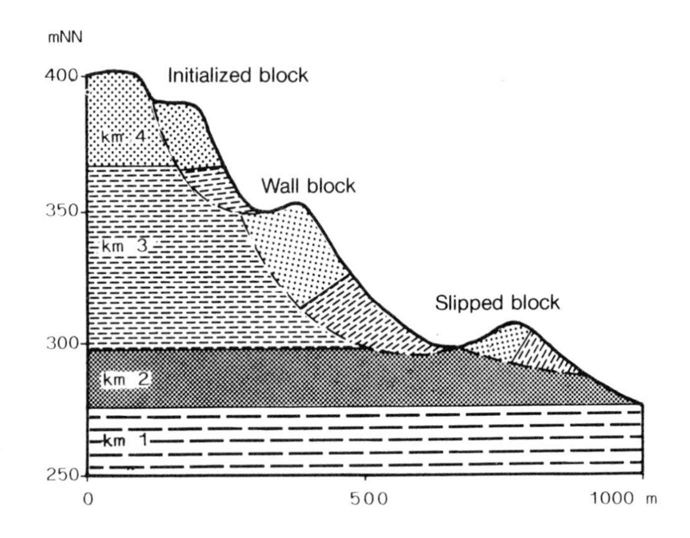

Fig. 6 Block glides on the Stromberg, Wurttemberg Keuper Hills (from Blume & Remmele 1989)

The widespread block slips on the Stromberg are the result of periglacial processes during the humid phases of the Pleistocene glacial epochs. Three main types of slip can be recognized:

1) Initial block - vertical movement is only several metres to tens of metres on a planar or slightly curved slip surface.
2) Wall block - rotational slip movement leading to a tilted block.
3) Slipped block - continued rotational slip leading to nearly 90° rotational angle; only located in lower slope area.

Owing to their longer transportation paths and their relatively older age, the slipped blocks are characterized by more mature forms. The slopes of the Stromberg are formed by a sequence of slightly dipping Keuper sediments, with varying resistances:

km 4: Stuben Sandstone;
km 3: Bunte Marl;
km 2: Schilf Sandstone;
km 1: Gipskeuper.

Surficial water flow

Surficial water flow during the intense precipitation events which are common in the seasonally humid tropics and subtropics, but which can also occur in semi-arid zones, is a strong erosive agent. Water flow on slopes occurs in rills and gullies (Photo 2.22). It is amplified by poor vegetation and it has become much more widespread as a result of anthropogenic effects (soil erosion; Photos 11.11 and 11.12). Linear erosion in gullies or small channels forms serrated slopes (Photo 2.23) or, in extreme cases, earth pyramids (Photo 2.24). Owing to the concentration of surface water flow into channels or gullies with concentrated erosion effects, triangular or V-shaped microforms develop, so-called chevrons (French: V-shaped uniform decoration; Photo 2.25). They are particularly evident in horizontally and slightly tilted structural landforms in arid zones.

Surface water flow in the form of sheet floods (or sheet or surface wash; Photo 2.26) is the main agency in the shaping and lowering of land surfaces. The strong erosional potential of sheet floods is because the pore volume of soils is filled with air after long dry periods; floodwaters during sudden, strong rainfalls move over an air cushion and cannot infiltrate the soils at first. In the seasonally humid tropics, strong rainfall, especially during the initial periods which follow dry seasons, effectively remove deeply weathered loose material (Photo 2.4). Meanwhile, weathering processes continue to attack the base of the weathered surface layers during wet periods. These concomitant processes of aggressive weathering and erosion lead to continual lowering of remnant land surfaces (Photo 5.18). In arid areas with periodic or episodic strong rainfall, the resultant high effective transport energies enable large amounts of physically weathered debris to be transported. Together with lateral stream corrasion (Photo 2.29), these processes form surfaces with steeper slopes than the remnant land surfaces (pediments; Photo 5.19) in a narrow margin at the base of high-relief areas.

Fluvial erosion

Streams and rivers are fundamental forces in surface relief formation. They transport weathered material which is supplied by denudation processes. The transported material itself provides an effective additional erosional tool on the base and flanks of river beds (basal and lateral stream corrasion). Streams and rivers also play a major role in creating sedimentary landforms by depositing transported material on valley floors (Photo 5.14), in alluvial fans (Photo 5.15), on levées (Fig. 13), and in deltas (Photo 5.17).The effective geomorphological force of a river is a function of the interaction of many different parameters. The material composition of the fluvial load (Chapter 5) is of prime importance. Additional controlling parameters are flow type, water volume and flow velocity, together with the amount and sorting of the load. With local exceptions, the general rule applies that erosive processes dominate in the upper reaches of rivers, while deposition dominates in the lower reaches. In addition, rivers can be perennial (constantly flowing) or periodic (seasonal) or episodically filled (flash floods).

Fluvial incision (downcutting) is certainly the most effective erosional process, but transport and erosion are closely related. The linear erosive force is not so much a function of the flowing water but is largely the result of the abrasive force exerted by the coarse material being transported on the bed of the river. Turbulence within the flowing water is a more important factor than flow velocity, and is apparent in stationary, vertical whirlpools which form potholes (Photo 2.27), or in horizontal, contra-rotating eddies which move the coarse bedload which deepens the river bed (Photo 2.28a, b). Lateral stream corrasion (Photo 2.29) is a further erosive force which is primarily exerted on the outer flanks of channel curves.

Geomorphological process zones as a function of climate

The multitude of geomorphological processes are controlled by endogenic factors such as rock type and geologic structure as well as by the primary exogenic factor - climate. The dependence of weathering and erosion processes on these factors has already been discussed in the description of the individual processes. The following is therefore a brief review of geomorphological processes as a function of climate.

Weathering

Physical weathering is mostly induced by strong diurnal temperature contrasts. It is therefore particularly intensive in the form of insolation weathering in arid, low-latitude zones, as frost weathering in subpolar climatic zones, and globally at high altitudes with frequent freeze-thaw cycles. Chemical weathering is the most important process in the humid tropical zones, owing to the perenially high temperatures and humidity which emphasize hydrolytic weathering processes. Hydration weathering dominates in the humid cool-temperate climatic zones, while cementation and duricrust formation are a widespread characteristic feature in arid zones.

Erosion

Fluvial incision, the most important erosional process, can be observed in all climatic zones, but is most intense in humid tropical areas. Slope erosion is the most pronounced process in subpolar and arid zones, while surface erosion (sheet wash) dominates in the seasonally humid tropics. Strong mass movement processes are involved in denudation in the humid tropics as well as in subpolar zones. Aeolian processes are particularly intensive in the subtropical and tropical arid zones, owing to the sparse vegetation. The humid areas of the cool-temperate climatic zones are characterized by moderate erosion rates, with the exception of solution processes (see Chapter 1).

The endogenic factors - lithological type and geologic structure - are only of secondary importance for erosional processes within the various zones of climate-dependent geomorphology. They do, however, lead to a considerable differentiation of the forms which result from the respective geomorphological processes (see Chapter 6). Climate not only determines the basic global zoning of geomorphological processes - it is also the most important of all of the factors which control the type and intensity of individual geomorphological processes.

Individual processes rarely act in isolation; several usually play a combined role in each climatic zone. The various processes can be of similar importance, or an individual process can dominate as an erosive force. Some climatic zones are characterized by perennial geomorphological activity, as for example in the humid tropics, while others show only seasonal periods of activity, for example in the seasonally humid tropics. In tropical and subtropical arid zones, geomorphological activity that is initiated by heavy rainfall only occurs episodically. A specific location in an extremely arid zone can be isolated from geomorphological activity associated with rainfall for many decades, although landforming aeolian activity will occur.

Table 3 summarises the broad relationship between landforms and climate.

Table 3 Zones with similar geomorphological process combinations, their predominant surface forms, and their associated climatic zones (after Hagedorn & Poser 1974 and Troll & Paffen 1964)

Zones with similar geomorphological process combinations	Predominant forms	Climatic zones
Strong physical weathering Frost-dynamic processes Intensive surface wash processes Intensive fluvial processes	Frost-patterned ground, pingos, debris fans, smooth slopes, serrations, V-shaped valleys with flood-plains	Subpolar climates Cool-temperate climates
Moderate chemical, slight physical weathering Moderate fluvial processes Other processes weak	Virtually only relict forms: fossil peneplains, V-shaped valleys with and without floodplains or river terraces, hollows	Humid areas of cool-temperate climates
Chemical and physical weathering Gravitational mass movements Intensive slope wash Periodically strong fluvial processes	Crusts, debris fans, pediments, serrations, torrents	Arid areas of cool-temperate climates Humid winter, dry summer areas of warm-temperate subtropics (Mediterranean climate,)
Strong physical weathering Episodic strong surface wash Episodic fluvial processes Intensive aeolian processes	Crusts, duricrusts, debris and rock hamada, pediments, glacis, playas, dry valleys (wadis), aeolian erosion (deflation basins, wind shaped rocks) and depositional forms (dunes)	Arid areas of the warm-temperate subtropics Arid areas of the tropics
Chemical and physical weathering Intensive surface or slope erosion	Peneplains with trough-shaped valleys, intermontane plains within mountain relief, inselbergs, V-shaped valleys with flood-plains, serrations	Humid-summer areas of the warm-temperate subtropics Seasonally humid tropics with low humidity
Strong chemical weathering Fluvial processes Surface wash Mass movements	V-shaped valleys, V-shaped valleys with flood-plains, peneplains and intermontane basins, slides, glides	Perennially humid areas of the warm-temperate subtropics Seasonally humid tropics with high humidity
Intensive chemical weathering Intensive fluvial processes Very strong mass movements	V-shaped valleys with flood-plains in low-relief areas, V-shaped valleys with steep slopes in high-relief areas, slips, slides, glides, sub-forest solifluction	Tropical wet climates

2.1 Insolation weathering
Repeated diurnal heating and cooling of rock surfaces – with associated volume expansions and contractions – induces tension between the exterior and interior rock zones. This loosens the intergranular cement, resulting in rock degradation or fracturing which leads to boulder or block formation.

near Cue, Western Australia

2.2 Frost weathering
Subzero temperatures cause ice formation in surficial fractures. The volume expansion during freezing forms and expands fractures, leading to rock destruction. More frequent freeze-thaw phases amplify the force of frost weathering.

near Ny Alesund, Spitsbergen

2.3 Exfoliation weathering
Exfoliation occurs in massive rock bodies as a result of pressure reduction which occurs when overlying rocks are removed by erosion. The exfoliated scales or shells can be up to several tens of metres thick, and can reach dimensions of several hundred metres. Dome or bell-shaped mountain forms are characteristic products of exfoliation weathering.

Pao de Acucar (Sugar Loaf), Rio de Janeiro, Brazil

2.4 Hydrolytic weathering
Granite blocks with 'fresh' cores are often preserved in the loose, ferrosiallitic, weathered residue that is produced by hydrolysis. The bleached colours indicate oxygen deficiency in the lower weathered parts. Zones closer to the surface are reddish-coloured due to the presence of iron-oxides.

near Antananarivo, Madagascar

2.5 Salt weathering
Rapid evaporation in arid zones draws saline solutions to the surface. Salt precipitation and the concomitant pressures caused by the growing crystals form flakes of weathered material.

Sandstone surface near Kayenta, Colorado Plateau, Arizona, USA

2.6 Exfoliation (desquamation)
A combination of salt and insolation weathering creates weathering shells with thicknesses of up to several tens of centimetres. Removal of the weathered shells is followed by renewed shell formation.

near Meekatharra, Western Australia

2.7 Crust formation

This mountain range, which consists of Pliocene calcareous sandstones, is covered by a calcareous crust (calcrete) formed by capillary water transport and surficial limestone precipitation. Recent erosion has partially destroyed the fossil crust, which was formed under less arid conditions during the Pleistocene.

Djebel Qarah, Eastern Province, Saudi Arabia

2.8 Duricrust formation

Duricrust formation is commonly explained by 'core weathering' processes, during which rock degradation proceeds from the subsurface to the surface of exposed rocks. It is a characteristic feature in arid climates. The friable, bleached material is removed from behind the partially overhanging duricrust remnants by wind or water. When duricrusts have been removed, renewed crustal formation commences.

near Mount Magnet, Western Australia

2.9 Tafoni formation

Caverns in rock walls and in large blocks in massive rock types are commonly described as tafoni. They are created by 'core weathering' processes with simultaneous duricrust formation, and by partial removal of the duricrusts. Canopy-shaped, overhanging cave roofs are a characteristic feature.

Les Calanches near Piana, west coast of Corsica (photograph by Ch. Iven, Rösrath, Germany)

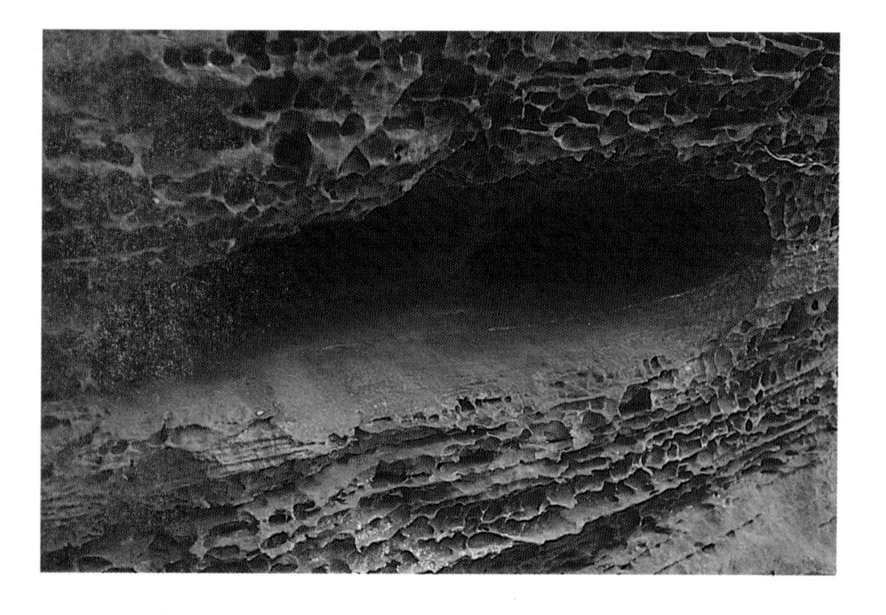

2.10 Seepage water weathering

Honeycomb or cell-shaped weathered structures are a typical feature. The thin, often hourglass-shaped separating walls are cemented areas, between which the sandy, friable material has been removed following solution weathering. Cigar-shaped cavities, in this example approximately 1 · 5 m across, can be correlated with impermeable, clay-rich layers. The surfaces of these layers form an iron-rich crust, above which the loose material is removed.

Luxembourg Sandstone near Echternach, Luxembourg

2.11 Talus slope

The talus slope is formed by several debris cones. The central cone is covered by fresh, coarse rock material. The debris mostly originates directly from rockfalls, rockfall channels, or from chimneys (large fissures).

Western slope of the Sella Group, Dolomites, Southern Tyrolia, Italy

2.12 Rock pillar

In horizontally-layered sandstones, tectonically-formed criss-crossing networks of vertical fractures form weakened zones which are preferentially attacked by weathering and erosion. The widening of these fractures leads to the formation of rectangular remnant rock pillars.

Prachovske skaly (City of Rocks) near Jicin, Czechoslovakia

2.13 Overhanging cliff
Highly resistant layers in a cliff are eroded more slowly than less resistant layers. They protrude from rock walls or form overhanging cliffs if they cap the exposed sequence. The example shown is formed by the highly resistant Main Conglomerate (*Hauptkonglomerat*) of the Triassic Bunter Sandstone.

La Chatte Pendue near Schirmeck, North Vosges, France

2.14 Mushroom rock
Mushroom or pedestal rocks are indicative of selective erosion by wind corrasion in arid areas. In the example shown, the base of the mushroom consists of soft marl, while the cap is formed of more resistant calcareous sandstones. In addition to wind corrasion, washouts in the marly base have helped to sculpture this microform. They can be seen in the gully in front of the mushroom and in the adjacent small channels. Average annual rainfall in this desert area is 80 mm.

Djebel Qarah, Eastern province, Saudi Arabia

2.15 Natural arch

This microform is created by selective weathering combined with removal of the weathered residue by wind and water. Similar processes lead to the formation of arches in sandstone cliffs. Some arches and bridges are, however, formed by fluvial erosion, for example when a meandering river cuts through the meander bluff before abandoning a loop. In the example shown, relict forms, which were created during the more humid periods of the Pleistocene, are preserved on the now semi-arid Colorado Plateau.

Arches National Park, Colorado Plateau, Utah, USA

2.16 Gelisolifluction

Gelisolifluction (or congelifluction) in subpolar climatic zones is a much stronger gravitational mass transport process than soil creep in humid cool-temperate climates. It occurs in the thawed upper layer of soil above the permafrost base. In the example, a very slight slope was sufficient to initiate microsolifluction movements which formed stone patterns (see Photo 5.20). Soil creep then extended these downslope into parallel or subparallel rows of rocks separated by finer material.

near Ny Ålesund, Spitsbergen

2.17 Rock fall

An example of rock falls on a scarp which is covered by weathered residue is shown in Photo 6.21, but rock falls occur in all climatic zones. They occur on steep slopes or cliffs when weathered parts of the rock face are detached from the unweathered bedrock. In the example shown, the semi-circular area can be seen in the rock wall from where the debris has fallen onto the talus cones. The large blocks in the foreground are from the upper part of the rock wall, as indicated by the horizontally emerging tree trunk. The fine-grained debris is removed by water flow.

near Greenvale, Queensland, Australia

2.18 Block glide

Mass transport in block glides (see also Fig. 6) often occurs when blocks of more resistant rocks move over a water-saturated base of impermeable clays. The example shows a block glide on a slope in the Swabian Alb area in Germany: two limestone blocks moved down extended fractures by several metres. Photo 6.11 shows an overview of a block glide which is associated with a rock fall.

Hirschkopf near Moessingen, Swabian Alb, Germany

2.19 Slip en masse

Slips occur in fine-grained weathered soils which can move en masse with only minor deformation. Wet soils are a prerequisite and the process is accelerated by the presence of an impermeable clay base which serves as a slip plane. In Central Europe, slips are mostly anthropogenically induced.

near Frickenhofen, Schwäbisch Gmünd, Württemberg Keuper area, Germany

2.20 Slip mass

This type of slip occurs when fine-grained weathered material swells due to water adsorption. The vegetation cover is disrupted and irregular slip processes occur. Small detachment scarps and bulges of accumulated material form a highly irregular, surficial microrelief.

Eichenhain, Stuttgart-Riedenberg, Germany

2.21 Debris flow

Sudden water saturation of debris or talus slopes in mountainous areas due to heavy rainfall or sudden snowmelts can turn temporarily deposited debris into a fluid, mud-rich debris flow. These flows tend to follow the normal erosional paths in stream and river beds. Characteristic features are debris walls on both sides of the flow bed, which are deposited due to slower flow rates at the sides of the flow. In the flat end reaches of the flows, large debris and mud cones are formed.

near Casaccia, Maira Valley, Val Bregaglia, Graubünden, Switzerland

2.22 Rill wash

Rill wash is a common phenomenon in areas with heavy rainfall. Closely spaced V-shaped channels (rills) on steep slopes are separated by sharp-edged ridges.

Inactive ash cone of the Bromo volcano, Tengger Mountains south of Surabaya, Java, Indonesia

2.23 Slope serration
In areas where easily erodable clay sediments dominate, rill washing of slopes forms an extensive network of serrations. The ridges between the rills are themselves cut by serrations which conduct water flow into the main channels. Strong solar radiation can form resistant crusts on the clays which often provide a protective coating against further erosion.

near Tainan, Taiwan

2.24 Earth pyramids
The microrelief of these steep cones is an extreme form of serration. In the example shown, earth pyramids have been sculpted out of a Pleistocene moraine. They are often capped by moraine boulders which provide temporary protection from sudden heavy rainfalls and continued erosion.

Southeast slope of the Ritten near Klobenstein, Southern Tyrolia, Italy

2.25 Chevrons
Surface water flow downslope can be concentrated into several major channels which form V-shaped, triangular erosional microforms. These chevrons are characteristic features of the shallow flanks of cuestas in arid zones. They can be stepped due to the varying resistance of the layers which form the slope.

Aures Mountains, Algeria

2.26 Sheet wash
Surface water flow resulting from severe rainfall is not only an effective erosional agent in the seasonally humid tropics but also in arid zones. Sheet wash surfaces are formed at the base of slopes, and large amounts of eroded material from the slopes are effectively transported by periodic or episodic sheet floods. These basal areas of slopes are called pediments (see Photo 5.19).

near Meekatharra, Western Australia

2.27 Pothole

Turbulent water flow is often marked by stationary, vertical whirlpools in which the coarse bedload erodes the channel floor and forms potholes. In the deep gorge, water-filled potholes indicate episodic water flow after heavy rainfall.

Black Mesa, Colorado Plateau, Arizona, USA

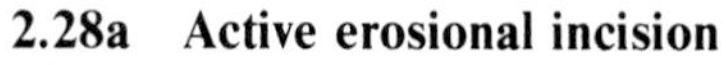

2.28a Active erosional incision

The approximately 20 m deep incision indicates active erosion which is, however, only episodic in this case. The photograph shows the uppermost section of a fluvial cut in a semiarid climatic zone. Water flow velocities are extremely high and flow is turbulent. Linear erosion is caused by the abrasive effect of coarse detritus transported in the river.

Haystack Buttes, South Dakota, USA

2.28b Inactive erosional incision

In areas of present extreme aridity, inactive erosional features are relict forms of previous active erosional processes. The cut shown in the example marks the upper boundary of an erosional zone which was active during a less arid phase. Recent inactivity, owing to the present extreme aridity, is indicated by the sand drifts as well as by the manganese crusts on boulders.

Djebel Ben Ghnema, Central Sahara, Libya

2.29 Lateral corrasion

Lateral stream erosion occurs by undercutting river banks, primarily on the outer side of meanders. It is a much more effective process in unconsolidated material than in solid rock, and in V-shaped valleys with flood-plains it leads to widening or lateral shifts of the river bed on the valley floor.

Roseg Valley, Upper Engadin, Switzerland

3 Tectonic landforms

Tectonic processes as a controlling factor

Every part of the earth's surface has been affected by tectonic movements during one or more phases of geologic history. Faulting and folding are the two most important tectonic processes.

Faulted structures

Tectonically-induced compressional or tensional forces can form fractures or cracks in rocks without causing actual tectonic movement or displacement. These fractures act as structurally weakened starting points for weathering processes. A closely-spaced network of fractures with different directions can lead to the development of specific typical landforms by weathering and erosion (rock pillars; Photo 2.12). Faults (Fig. 7) are fractures along which movement or displacement is occurring or has occurred in the geologic past as a result of compressional or tensional forces. As fault movements are rarely gradual but mostly occur at irregular intervals in time and at various locations on the faults, they are often marked by earthquakes. Downthrown fault planes, which indicate tensional processes within the earth's crust, are angled down to the downthrown block or fault trough. Thrusting movements, resulting from compressional forces, occur along fault planes which dip down below the upthrusted block. If the fault angle is less than 45°, the movement is defined as overthrusting. Downthrown movements are indicative of crustal extension, whereas up- and overthrusting movements involve a reduction of the crustal surface area. Fault movements can move a single bed into different elevations along a fault plane. Movements can occur along angled as well as vertical fault planes, and can become morphologically evident as steps or fault scarps (Photo 3.1). If a fault zone instead of a single fault marks the boundary between two blocks, the resultant morphology can be a steplike (*en echelon*) series of fault scarps (Photo 3.2). A block between two faults which has been thrust upwards relative to the adjacent blocks is commonly called a horst. If the horst has only been uplifted on one side, a tilted fault block results (Photo 3.1). A graben (from the German: ditch) is an elongate, relatively narrow crustal segment which has moved down between two adjacent blocks (Photo 3.2).

The total length of graben faults and their throw (i.e. vertical displacement) can differ enormously. The great graben systems of the world are characterized by considerable throw distances: for example, the depth of the Upper Rhine Valley Graben in southern Germany exceeds 3000 m (Fig. 8), while the Red Sea Graben has been depressed by more than 6000 m. Both of these graben represent subsections of continental graben or rift systems. They are bounded by faults and are the result of tensional forces within the earth's crust leading to crustal expansion. The origin of these giant graben (taphrogeny) is explained by the introduction of ultrabasic material from the upper mantle, which causes mantle bulging, and the process is regarded as the initial stage of continental rifting (see Chapter 1). Elevation differences between blocks created during graben formation and the concomitant increase in relief energy initiate powerful morphodynamic processes. Horst blocks become zones of intensive erosion which removes large volumes of

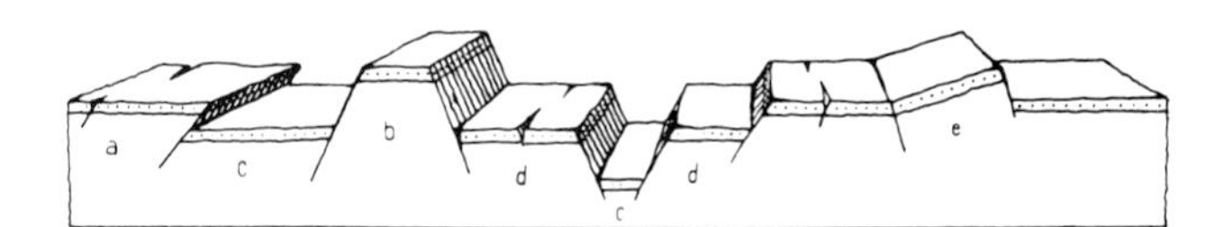

Fig. 7 Basic block faulting types (from Louis & Fischer 1979)

a: Horst = relatively uplifted block, bounded by upthrow faults (compressional faulting)
b: Horst = relatively uplifted block, bounded by downthrow faults (tensional faulting)
c: Graben = relatively depressed block, bounded in this case by *en echelon* downthrown faults
d: *En echelon* blocks, bounded by *en echelon* faults
e: Tilted fault block, tilt block

The vertical elevation differences of the dotted layer represent the vertical throw of the respective fault.

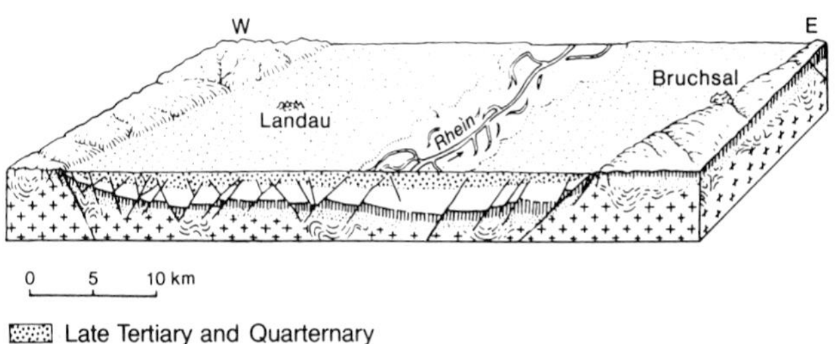

Late Tertiary and Quarternary
Early Tertiary
Muschelkalk and sections of Keuper, Lias, and Dogger
Rotliegendes and Buntsandstein (Permian/Lower Triassic)
Variscan basement

Fig. 8 Tectonic section of the Upper Rhine Valley Graben (from M. & D. Richter 1981, after J. H. Illies)

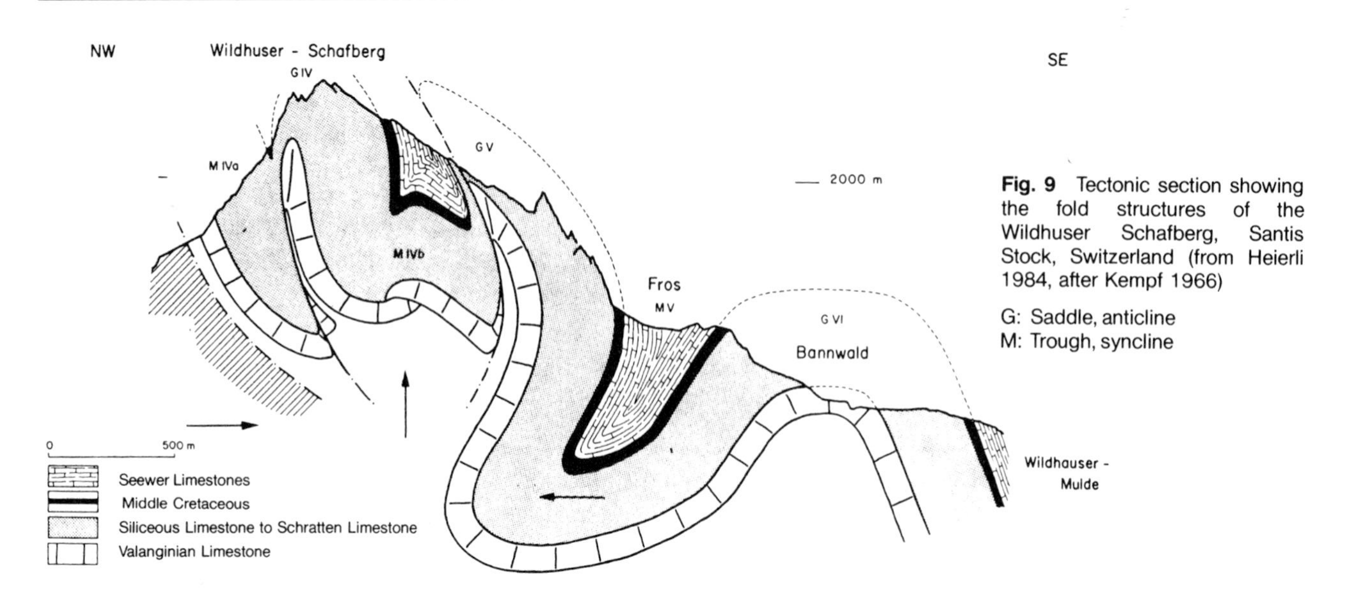

Fig. 9 Tectonic section showing the fold structures of the Wildhuser Schafberg, Santis Stock, Switzerland (from Heierli 1984, after Kempf 1966)

G: Saddle, anticline
M: Trough, syncline

rock and forms typical dissected scarps (Photo 3.2). The graben itself is filled with corresponding amounts of erosional debris (Fig. 8), a process which can conceal the enormous throw distances of the graben faults and can prevent their morphological expression as fault scarps. Small, secondary graben systems with moderate throw distances frequently accompany the continental systems and provide additional evidence of tensional forces in the crust.

Fault scarps also exist which mark boundaries of horizontal block movements. Large-scale horizontal movements along fault planes, as for example along the San Andreas Fault in California (Photo 3.4), can indicate horizontal movements of lithospheric plates along transform faults (see Chapter 1).

Fold structures

Fold structures are created by compressional forces within the earth's crust and are most evident when the rocks are clearly layered. Flexures represent transitional features between faulted and folded structures and are characterized by layer bending without faulting. However, many faulted structures have evolved from flexures. Folds can be subdivided into high areas called anticlines and low areas called synclines. The size of the folds can vary widely and is controlled by the type and thickness of a rock sequence as well as by the intensity of the folding process, i.e. by the amount of shortening of the section. Extreme compression can lead from simple folding to tipped and recumbent folding and finally to overthrust folds. Horizontal forces can be so extreme that entire layer sequences can be thrust over the underlying foreland by up to several tens of kilometres. The overthrust sheets themselves can be folded again (folded thrust sheets; Fig. 9, Photo 3.5). Thrust sheet formation results in considerable section shortening, for example up to 66 per cent in northern Switzerland. In conditions where the sediment sequences are only slightly folded, as for example in the marginal areas of fold belts, simple anticlines (Photo 3.6) and synclines can be recognized. The angle of sedimentary beds often affects their relief forms: low-angle beds are associated with tabular mesas and buttes, whereas tilted beds are associated with asymmetrical, low-angle ridges (cuestas) or symmetrical, high-angle ridges (hogbacks) (see Chapter 6).

Structural types of continental surface forms

The aforementioned tectonic processes are responsible for the generation of the main structural types of surface relief. As already mentioned in Chapter 1, two large-scale morphotectonic units can be recognized in the continental relief of the earth's surface: continental platforms and mountain chains. Their basic geological structures are summarized in the following sections.

Mountain chains

The structures of these morphotectonic units are the result of folding processes in convergent zones of colliding plates. Mountain chains can be classified into three structural types according to the general fold intensity: simple anticlinal and synclinal types, fold belts, and thrust belts. Their mountainous character and elongate forms are outstanding features of the present mainland relief (see Chapter 1). However, their high elevation and rugged character is not just a direct function of their folding, but mainly reflects the recent tectonic uplift that occurred after the multiphase folding processes. Rapid uplift then created the high relief energies which are a prerequisite for intensive geomorphological processes in the mountain chains. Volcanic and seismic activity is an additional characteristic feature of mountain chains: earthquakes can trigger spontaneous and intensive geomorphological processes in the form of large-scale rockfalls and landslides. The main fold and uplift processes that formed the present mountain chains occurred between the Cretaceous and the Late Tertiary, and uplift has persisted to the present time. This youngest, so-called Alpine, mountain-forming phase was, however, preceded by other widespread phases of tectonic activity: for example in Europe, the Variscan (mainly during the Devonian) and the Caledonian phases (mainly during the Silurian). Areas which were affected by these early mountain-forming phases and which mostly no longer retain their mountainous character have accreted to the margins of continental core areas and have long since become integral components of the continental platforms.

Continental platforms

The oldest and most stable parts of the earth's crust are the continental shield areas. These consist mainly of folded rocks of pre-Cambrian age which have since been eroded. Examples are the Baltic and Canadian shields. They are mostly slightly domed (craton) and occasionally, especially in Southern Africa, subdivided by rises. Remnant land surfaces determine their relief. Their margins are mostly buried beneath discordant layers of sedimentary rocks of either marine or terrestrial origin which were deposited in environments essentially similar to the present continental shelf zones. Owing to the absence of fold structures and the scarcity of fault structures, these areas are defined as the stable shelves. Examples occur on the Russian and Siberian shields, where horizontally layered sediments form a large-scale, tabular relief. In contrast, unstable shelf areas of the continental platforms are zones which accreted to the continents during the previously mentioned post-Cambrian tectonic phases and were since subjected to various types of fault tectonics. This structural type is therefore also described as a fault block tectonic type. Regional examples are the Mittelgebirge (highland areas) in Western and Central Europe. Characteristic features are a basic structure composed of a mosaic of uplifted, downthrown, or tilted blocks. Tilting of the cover rocks, which mostly consist of discordant layers of Mesozoic sediments, has formed the present relief of flat and sloping surfaces. In addition to the abundant fault systems, with varying throws, which are responsible for the block mosaic structure, the unstable continental shelf also features extensive graben systems as exemplified by the East African and Near Eastern rift systems, which are interpreted as initial stages of continental rifting. The continental platforms also include extensive alluvial plains as exemplified by parts of the North German Plain, Mesopotamia, or the North American Gulf Coast area. These alluvial plains are included in the recent continental shelf areas. They are areas of tectonic subsidence, with alternating sequences of terrestrial and marine sediments which are still being deposited at present.

3.1 Fault scarp

A fault scarp separates an uplifted from a depressed block. The eastern edge of the Teton Range in the background is one of the major North American fault scarps. The Teton block, which reaches its maximum altitude in the Grand Teton (4196 m), is a tilted block separated from the downthrown, sediment-filled Jackson Hole in the foreground by a nearly 50-km-long, linear fault with an estimated vertical throw of about 6000 m. Along the fault scarp, the Teton Range is formed by Precambrian magmatites and metamorphites. Paleozoic sediments cover the western part of the range and the tilt of the block is indicated by the westward dip of these sediments.

Teton Range from the east, Wyoming, USA

3.2 En echelon faults

The tensional tectonic forces which are instrumental in forming grabens often generate marginal sets of elongate fault blocks. Fault movements occur along a parallel sequence of fault steps, so-called *en-echelon* faults. The boundary zones between horst and graben areas are therefore subdivided into several increasingly depressed blocks, which are morphologically expressed by scarp steps. This can be observed for example in the Upper Rhine Graben, one of the main European graben systems. Erosive dissection of the fault scarp is strong, due to the considerable vertical throw of the graben faults.

Western margin of the Black Forest, north of Bühl, Baden, Germany (photo: Landesbildstelle Baden, Karlsruhe)

3.3 Graben

The Valle del Cibao Graben, on the Caribbean island of Hispaniola, has a length of 250 km and a maximum width of 40 km, and is roughly the same dimensions as the Upper Rhine Valley Graben in Germany. Sediment thicknesses of 10 000 m on the depressed block indicate the extreme vertical displacement of this graben structure between the Central and Northern Cordillera of Hispaniola.

View from the Central Cordillera over the Valle del Cibao towards the Northern Cordillera, east of Santiago, Dominican Republic

3.4 Horizontal fault movement

Transform faults are strike-slip faults along which large-scale horizontal block or crustal plate movements occur. The Pacific Plate, for example, is separated from the American Plate by the approximately 600-km-long San Andreas fault and drifts northwards at a rate of about 6 cm per year. This process is not a gradual, constant movement, but occurs intermittently as indicated by earthquakes. For example, during the Loma Prieta earthquake (magnitude 7 · 1 on the Richter scale) on 17 October 1989, the relative plate movement was 2 metres. In the Santa Cruz Mountains to the south of San Francisco, the Pacific Plate moved about 2 metres in a northwesterly direction and was simultaneously thrust on to the American Plate by about 1 · 5 metres. The aerial photograph clearly shows the linearity of the San Andreas fault. Recent activity is indicated by the south-easterly deflection of valleys which cross the fault at right angles.

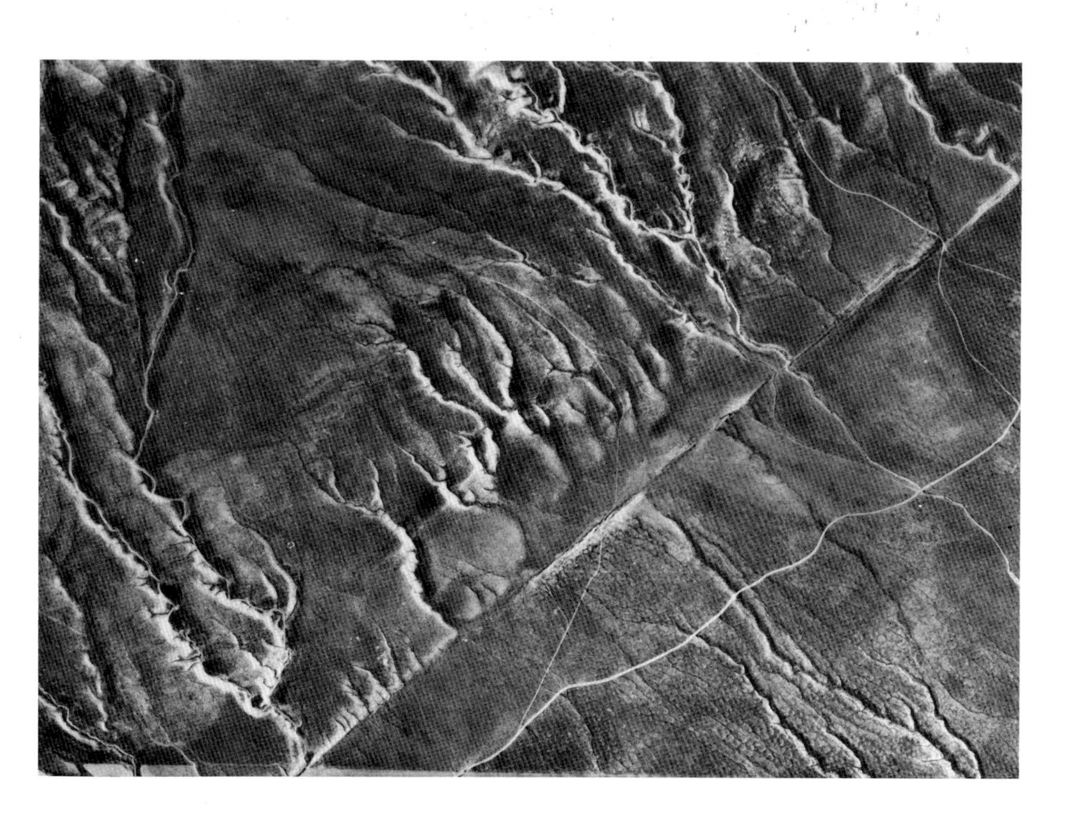

(photo: USGS; from Kronberg 1984)

3.5 Folds in thrust sheets

The geological structure of this section of the alpine Säntis Stock is also shown in Fig. 9. The Säntis thrust sheet, which overlies the Wildhuser Schafberg, is one of the Helvetic thrust sheets which have been overthrust in a northwesterly direction as a result of strong crustal shortening processes. The Wildhuser Schafberg Anticline (G IV in Fig. 9) is one of six parallel, southwest–northeast trending saddles which have been eroded into ridges to form the relief of the Säntis Stock. These folds are mostly recumbent in the thrust direction and dissected by transverse faults. The erosional competence of the Cretaceous rocks of the folded thrust sheets varies widely. The karstic Lower Schratten Limestone is particularly resistant and forms the ridges, e.g. of the Wildhuser Schafberg and the Zehenspitz on its southeastern flank. Less resistant layers, such as the underlying Drusberg marls, have been removed by erosion. Relief forms are a direct function of lithological types.

Wildhuser Schafberg, Säntis Stock, Switzerland (photo by K. -G. Krauter, Esslingen)

3.6 Anticlinal structure

In anticlines, sedimentary rocks are forced into dome- or saddle-shaped structures. Selective weathering and erosion of layers with varying resistance creates tabular landforms (mesa, butte), or if the dip angle of the layers is steeper, *en echelon* ridges (cuesta, gohback) are formed. This relief, however, only partially surrounds the dome-shaped structure shown here in the middle of the photograph. In the foreground the anticlinal structure is interrupted by an additional anticline, which is also morphologically evident as a series of parallel ridges The overlapping anticlinal structures were formed by fault block tectonics which is a characteristic feature of the marginal areas of folded crustal zones. The anticlinal structures have then been sculpted by exogenic processes.

Zagros Mountains south of Schiras, Iran

4 Volcanic landforms

Plutonism and volcanism

Volcanism or volcanic processes are those in which magma, i.e. molten rock material and gases, rises from deeper levels of the earth's crust and from the mantle. They can be subdivided into deep volcanic processes and those that reach the surface. Deep processes in which magma does not reach the surface are described as plutonic. Plutonites are formed when magma solidifies or crystallizes within the crust. These large bodies or plutons of intruded igneous material are called batholiths if they have solidified at great depth as dome-shaped magma complexes of typically large areal extent. Smaller igneous masses at shallower depths and with less regular shapes, which result from intrusions into layered rock sequences, are defined as laccoliths. Intrusive masses such as granites, which are uncovered by erosion of the overlying rocks, often become geomorphologically evident as volcanic stocks or plugs owing to their higher resistance (Photo 6.6).

Surface volcanism occurs when rising magma is extruded on to the earth's surface. The molten magma is ejected as lava which is identical in composition to the magma except for the gaseous components which have escaped during eruption. In addition to the extrusion of lava, loose volcanic debris or ash (pyroclastics or tephra) is also commonly ejected. Surface volcanism, which is discussed in the following sections, forms a suite of specific, readily identifiable surface forms. The wide variety of forms is a consequence of the different volcanic processes, as well as subsequent geomorphological processes.

Volcanic surface structures

Volcanic products

The most important reason for the variations in volcanic processes and the resultant forms are the viscosity and gas contents of the extruded material, which are a function of the temperature and chemistry of the magma. The decisive factor is the silica (SiO_2) content which enables the definition of three main magma types and therefore of three corresponding groups of igneous extrusives (volcanites):

1. basaltic rocks, the products of basic melts (< 52 per cent SiO_2);
2. andesitic rocks produced from intermediate (52–65 per cent SiO_2) and
3. rhyolithic rocks from acidic magmas (> 65 per cent SiO_2).

Basic lavas, such as basalt, are mostly non-viscous; their extrusion temperatures are high (1100–1200°C), gas contents escape rapidly, and lava flow velocities are high because of their low viscosity. On the other hand the gas-rich intermediate and acidic lavas are sticky or viscous and move slowly. Their eruption temperatures are approximately 200–300°C lower than basic lavas and their high viscosity hinders gas expulsion. This can generate severe overpressuring in the rising gas-rich magmas and subsequent catastrophic explosions, whereas the emergence of basic lavas mostly proceeds as an effusive process.

Concerning the resultant microrelief of the surface of solidified lava flows, two lava types are commonly defined by expressions originally used on the islands of Hawaii. Pahoehoe-type lava (Photo 4.1) is typical for predominantly basic lavas with low viscosities and low gas contents, while Aa-type lava (Photo 4.2) represents mostly intermediate and acidic, gas-rich, and viscous lava. Basic lava frequently solidifies into hexagonal (six-sided) columns (Photos 4.3, 4.14a) owing to contractive fractures that develop during cooling. Mineral structures and cooling rates also play a role in their formation.

The ejection of loose, pyroclastic material during explosive volcanic eruptions is a typical feature of volcanoes with intermediate and acidic magma supplies. The pyroclasts are mainly composed of lava fragments, together with rock fragments that have been torn from the sides of the feeder channels by the rising magma. Definitions differ, but particles are generally subdivided according to their size into slag, lapilli, and ash. Slag is formed by large lava fragments which are already partially solidified before being deposited on the surface. Lava bombs are, strictly, viscous pyroclasts which are ejected to great heights and solidify into ball-shaped clasts in flight. The expression 'lava bombs' is, however, often used for all large rock fragments. Lapilli are small, roughly pea-sized pyroclastic fragments, while ash represents the finest class of lava fragments.

Volcanic surface forms

The effusion of non-viscous lava from deep-reaching fissures (linear volcanoes) forms extensive basalt plateaux. These effusive sheets can reach thicknesses of 3000 m and cover considerable surface areas. The Parana Basalts of Brazil are the largest basalt plateaux of this type with a total surface area of more than 1 million km². The basaltic flows of the Deccan Plateau cover approximately 500 000 km² in India, while in North America the Columbia Plateau reaches 400 000 km². The largest plateau of this kind in Europe is the Vogelsberg in Germany with a total area of 2500 km². In all these areas the plateaux are built up by successive horizontal layers or flows of basalt, giving rise to a characteristic tabular relief. Their layered structure is frequently clearly exposed in erosive cuttings and is especially clear when volcanic ash has been deposited between the layers of basaltic material, as for example in the Western Ghats of India (Photo 6.10).

If the effusion of low-viscosity fluid lava occurs from a single central vent or feeder pipe, large shield volcanoes (Photo 4.4) are formed. As in the case of basalt plateaux, these result from repeated lava flows which create a sequence of stacked basaltic layers (Fig. 10a). Cone-shaped volcanoes are generally associated with intermediate or acidic lavas. They are described as stratovolcanoes if their internal structure is layered. They can consist entirely of pyroclastics (Photos 4.5, 4.6) or of alternating layers of lava and pyroclastics (Photo 4.7). The pyroclastic type of volcano usually seems to achieve the closest resemblance to the classical cone shape. Parasitic cones (Photo 4.8) are secondary craters which occur on the side of larger volcanic structures, and can also be indicated by bulges on volcano flanks. Their feeder conduits can be branches of the main feeder or can come directly from the main magma chamber. Volcanic domes or stocks (Photo 4.9) are transitional forms between volcanism and plutonism. They are dome-shaped lava bodies with steep flanks, which are formed by the slow upward movement of viscous, semi-solidified lava. If the lava has already solidified in the feeder conduit, it can be pushed out to form a needle-shaped volcanic plug.

Hollow forms

Lava flows contain gases which can form cavities of various sizes in the solidified lava. Lava tunnels can appear in low-viscosity basic lavas when molten lava flows beneath an already solidified lava surface (Photo 4.10). The most impressive hollow or concave volcanic forms are the craters of volcanic structures. Craters form as cone- or bowl-shaped depressions around the upper end of feeder pipes. Stratovolcanoes are typically capped by funnel-shaped craters (Photos 4.5, 4.6, 4.10, 4.11) that are modified by each eruption. Craters on shield volcanoes are usually kettle-shaped collapse or pit craters (Photo 4.12) with steep or even vertical walls. They resemble calderas, more or less circular, steep-walled volcanic depressions with diameters of 5–20 km, which are formed by collapse processes or by explosions (Fig. 10c, Photo 4.12). Steam eruptions, caused by the explosive evaporation of ground water on contact with magmatic material, can blast out funnel- or bowl-shaped craters in non-volcanic (country) rocks which are then surrounded by crater rings of country rock fragments. They are called maare from their type occurrence in the Eifel hills of Germany (Photo 4.13).

Remnant volcanoes

Volcanoes are exposed to erosional processes as soon

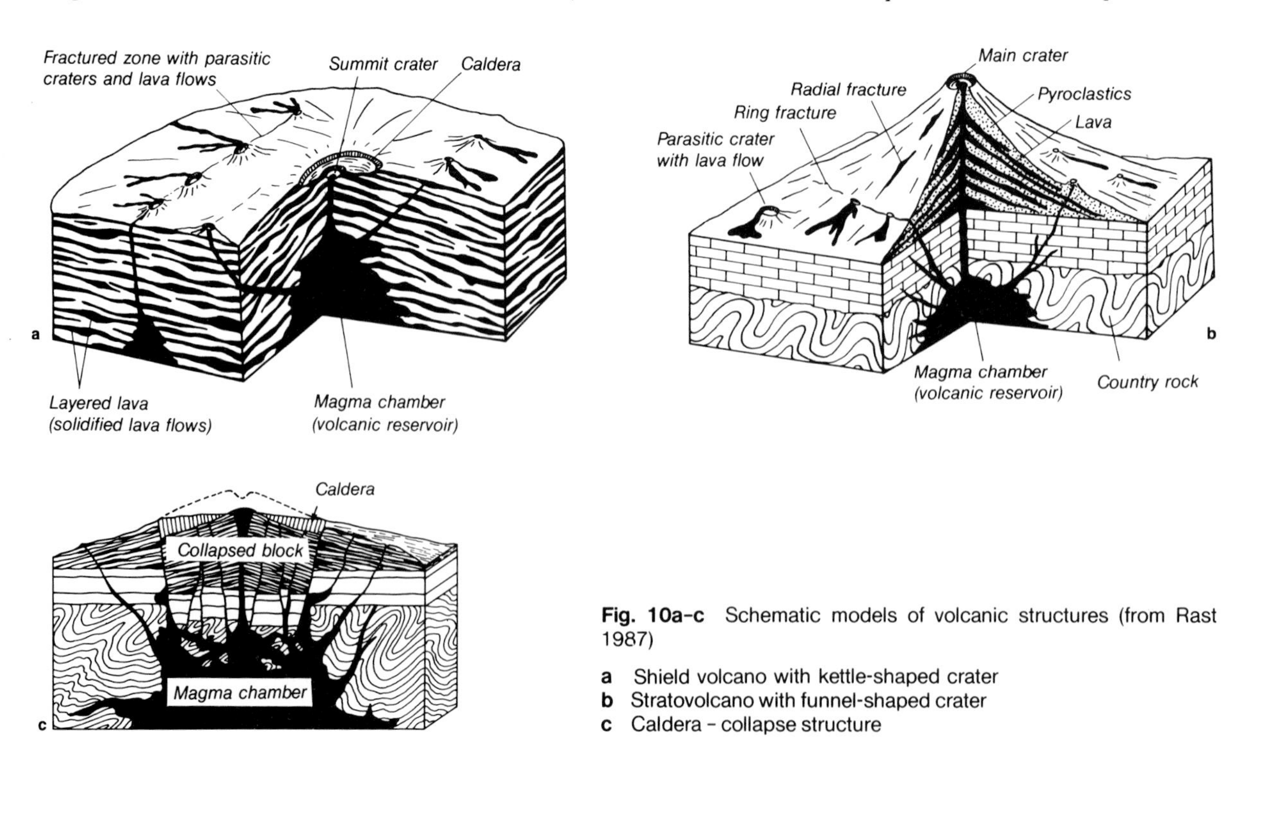

Fig. 10a-c Schematic models of volcanic structures (from Rast 1987)

a Shield volcano with kettle-shaped crater
b Stratovolcano with funnel-shaped crater
c Caldera - collapse structure

as they are created. The effects of geomorphological processses on volcanic structures have already been shown in several examples (Photos 4.4 to 4.9). Remnant volcanoes are created when selective erosional processes isolate the magmatic filling of the volcanic conduits. The higher resistance of the volcanic rocks in the conduits preserves them, while the surrounding pyroclastics are removed (Photos 4.14a, 4.14b).

Post-volcanic processes

Exhalations of gas and water vapour from fumaroles at temperatures of 200 to 900°C, from solfatare at 150 to 400°C, or in mofettes (cool carbon-dioxide exhalations) have virtually no effect on relief-forming processes. However geysers, which periodically eject hot water and steam (Photo 4.15), are often surrounded by characteristic cones and terraces of siliceous sinter.

Distribution of active volcanism

Presently active volcanism shows a clear distribution pattern. It is related to zones in which new oceanic crust is formed in the mid-oceanic ridges (oceanic rift volcanism), to active continental margins and island arcs (subduction zone volcanism), to continental graben systems (continental rift volcanism), and to hot spots (oceanic intra-plate volcanism). This spatial differentiation of volcanic activity as a result of plate-tectonic processes corresponds to a differentiation of the products, mechanisms, and subsequently of the forms of volcanic structures.

In the mainland areas of the earth's surface, forms that are directly related to active volcanism are restricted to relatively limited areas. Continental platforms, one of the two large-scale morphotectonic units of the continental areas, do not show active volcanism, except for the graben systems of the unstable shelf areas. However mountain chain areas do show intensive volcanic activity (see Chapter 3).

Most of the continental volcanic activity occurs on the active continental margins, which are morphologically marked by mountain chains and island arcs (see Chapter 1, Fig. 2). This volcanism is driven by subduction processes and includes the volcanoes of the circum-pacific region, of the Lesser Antilles, and of the Mediterranean. They produce intermediate and acidic igneous material derived from the melting of subducted continental crust. While most of these stratovolcanoes were created by explosive as well as by effusive eruptions and therefore show irregularly formed cones, some show evidence of extreme explosive activities. These consist almost exclusively of pyroclastic material and can therefore be recognized by their perfect cone shapes.

Continental rift volcanism is less widespread, as it is restricted to the continental graben systems (rift systems, see Chapter 1). The products show considerable variation as the initial development phase is characterized by basalt plateaux, while later production of intermediate and acidic magmas is derived from the melting of continental crust. Phonolithic and trachytic lava and pyroclastics are predominantly produced by a mixture of effusive and explosive activities. Volcanic landforms are therefore typified by cone-shaped stratovolcanoes in addition to the older basalt plateaux.

The volcanism of the mid-oceanic ridges only appears above sea level in a few locations, such as on Iceland and the Azores. Volcanic products consist mainly of tholeiitic-basaltic magma, in the form of either igneous sheets that spread out from linear volcanic fissures or of layered shield structures that are extruded from centrally located volcanoes.

Hot spots (see Chapter 1), which represent intraplate volcanic activity, appear in the form of oceanic islands. The predominantly effusive production of tholeiitic and alkali-basaltic magma from central volcanoes leads to the formation of huge shield volcanoes.

4.1 Pahoehoe-type lava flow (ropy lava)
A viscous skin forms on the surface of low-viscosity, gas-poor, basic lava flows, which can assume many different shapes when flow velocities are reduced. The term 'ropy lava' describes their most common shape. In the case shown, fountains of basic lava are ejected from several small spatter cones, so-called Hornitos, by the pressure of escaping gases (eruption date: 15 October 1970).

Mauna Ulu, eastern fracture zone of Mount Kilauea, Island of Hawaii

4.2 Aa-type lava flow (block lava)
This lava type is associated with slow, gas-rich lava flows with thick crusts. Continuing flow movements beneath the solidified crust tend to break up the crust and sharp-edged blocks of various sizes are formed: their rough, slag-like surface is explained by degassing processes. A block lava flow is shown from the Paricutin volcano which erupted in 1943 and remained active to 1952. The extruded lava covers an area of 25 km^2 and buried two villages during the first two years of activity. An intermediate melt, which solidified to form andesite, was initially ejected with a silica content of 55 per cent, which later rose to 60 per cent.

Parícutin, Michoacán, Mexico

4.3 Basalt columns
The formation of polygonal, mostly hexagonal, columns is a characteristic feature of many basaltic lavas. Their form is derived from fracture networks that develop in the lava as a result of contraction during cooling. The fractures and the resulting columns are aligned vertically to the cooling surface. An example from Northern Ireland is shown in which abrasive wave action has formed an extensive pavement of vertical, polygonal, basalt columns at median sea level.

Giant's Causeway, County Antrim, Northern Ireland (photograph by G. Remmele, Heilbronn)

4.4 Shield volcano

This volcanic structural type is created by repeated flows of basic, low-viscosity, lava that spread out from a well-defined, central effusion point. The flanks of shield volcanoes have low angles (3–6°), and their basal diameters can therefore exceed the total height of the structure by a factor of twenty. The island of Hawaii, with its two most prominent peaks Mauna Kea (4214 m) and Mauna Loa (4196 m), is regarded as a single shield volcano with several eruption locations. The basal diameter of this huge volcanic structure is approximately 200 km at a depth below sea level of about 4000 m, and its total height exceeds 8000 m.

Mauna Kea, view from the east, Island of Hawaii

4.5 Ash cones

Ash cones are created by explosive volcanic eruptions and consist entirely of pyroclastic material. Mount Bromo (2329 m) in the Tengger Mountains of Java is formed by two cones that consist entirely of ash, lapilli, and lava bombs, and rise from the ash-covered base of a caldera. The left-hand cone is active, the other inactive (dormant). A typical feature of ash cones is their steeply angled flanks; the permeable pyroclastics retain sufficient slope stability even at angles of 35°.

Mount Bromo, Tengger Mountains south of Surabaja, Java, Indonesia

4.6 Stratovolcano (pyroclastic type)

Volcanoes which are formed entirely of pyroclastics are relatively rare. Mount Quill (601 m) is a young, extinct ash volcano. As the grain size of the pyroclastic fragments decreases with increasing distance from the crater, the surface of the cone is angled at 40° near the summit and then gradually flattens out in a concave slope profile which is typical for all pyroclastic volcanoes. Two small andesitic domes are buried in the pyroclastic material which is of intermediate composition. One of these domes has pushed up the country rocks (Pleistocene White Wall limestones) on the southern flank of the volcano so that they dip seawards at an angle of 40° (right foreground). The steep, funnel-shaped crater has a diameter of 600 m and a depth of more than 300 m. As the activity of the Pleistocene volcano ceased only a few hundred years ago, the crater is well preserved and the erosional dissection of the flanks is relatively insignificant.

Mount Quill, Sint Eustatius, Netherlands Antilles

4.7 Stratovolcano (composite type)

Composite volcanoes are formed by lava flows as well as by pyroclastics and are therefore indicative of mixed effusive and explosive activity. This is common when intermediate magmas are extruded. Composite volcanoes and purely lava volcanoes rarely show perfect cone shapes. The volcanic cone of Mount Demawend (5670 m), built by an alternating sequence of andesitic igneous rocks, ash, and slag, towers as an isolated, younger volcanic structure about 1500 m above the peak level of the Elburs Mountains. These belong to the Near Eastern mountain belt and consist mainly of folded Jurassic and Cretaceous limestones and granitic igneous rocks.

Mount Demawend, Elburs Mountains, Iran

4.8 Parasitic cones

Large volcanic structures often carry secondary cones that can be fed by conduits branching from the main feeder or coming directly from the magma chamber. Mount Misery (1131 m) is a large, extinct stratovolcano that was formed during the Pleistocene by andesites and associated pyroclastics. Several craterless parasitic cones are located on its flanks, two of which are visible in the photograph. They are mostly sited above tectonic zones of weakness. The andesitic lava of the lower cone, Brimstone Hill, has distorted the Pleistocene limestone beds, resulting in similar features to those of Mount Quill (see Photo 4.6).

Mount Misery, Saint Kitts, Lesser Antilles

4.9 Volcanic dome

Domes or plugs are craterless volcanic structures formed by upward-moving, highly viscous, semi-solidified masses of intermediate or acidic lava, and they can often be identified by their characteristic steep walls. The two Pitons (Gros Piton 777 m, Petit Piton 732 m), which consist of intermediate lava (dacite), are easily recognizable landmarks on the Caribbean island of Saint Lucia. They are located in a large caldera together with thirteen other andesitic domes and seven small craters. The caldera itself is covered with pyroclastic debris and its edges have been almost completely eroded.

Pitons, Saint Lucia, Lesser Antilles

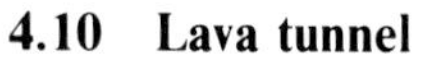

4.10 Lava tunnel

Lava tunnels can be created in low-viscosity basic lavas when molten lava flows beneath an already solidified lava surface. In the example shown, the roof of the lava tunnel has partially collapsed. The lava tunnel and the asymmetrical slag cones were formed by eruptions between 1730 and 1736 in the southwestern part of the Island of Lanzarote.

Parque Nacional de Timanfaya, Lanzarote, Canary Islands (photograph by Ch. Iven, Rösrath, Germany)

4.11 Funnel-shaped crater

This type of crater is a typical feature of stratovolcanoes (see also Photo 4.6). The example shown here is the crater of an ash cone. The layering of the fine-grained pyroclastic material is clearly visible on the crater wall, as are the lava bombs on the lower parts of the slopes. Gas exhalations continue in the deepest part of the crater between the active phases. The form of the crater is changed by each eruption.

Mount Bromo, Tengger Mountains south of Surabaja, Java, Indonesia

4.12 Kettle-shaped crater (pit crater)

Kettle-shaped craters with diameters of 100 to 2000 m are characteristic features of shield volcanoes. On the right hand side of the photograph, the steep sidewall of a caldera (diameter: 3 × 4 km) can be seen. Solidified pahoehoe-type lava covers its floor. This caldera contains Mount Kilauea's main crater, Halemaumau, a typical pit crater with nearly vertical walls. Until 1924, Halemaumau was a lava lake with intermittently rising and falling lava levels and constant intermingling of molten and semi-solidified lava. After the disappearance of the lava lake, heated ground water triggered a phreatic explosion that increased the diameter of the Halemaumau crater from 350 m to more than 900 m. Since then the main eruptive zone of Mount Kilauea has migrated to a fractured zone located further to the east.

Halemaumau, Kilauea, Island of Hawaii

4.13 Maar

Maars (German plur.: *Maare*) are funnel- or bowl-shaped, circular cavities which were created by gas explosions from secondary volcanic magma chambers: an explosive evaporation of ground water blasted off the cover rocks. The maars are mostly only several hundred metres wide and are usually surrounded by crater walls of country rock and volcanic material. In the humid climatic zone where they typically occur, in the Eifel hills of Germany, they mostly contain lakes. The example shown is one of twelve explosion craters in the western part of the island of Nosi-Be in northwest Madagascar.

Nosi-Be, Madagascar

4.14a Remnant volcano (basaltic conduit filling)

Conduits or feeder-pipe fillings of former volcanic cones are often revealed by differential erosion. The example shows a basaltic pipe with typical polygonal columns (compare with Photo 4.3) which assume more vertical positions in the higher levels of the pipe, in accordance with the position of the cooling surface.

Eastern slope of the Yemen Highlands between Sanaa and Marib, close to the old Marib road, Yemen Al-Yaman (photography by G. Remmele, Heilbronn, Germany)

4.14b Remnant volcano (asymmetrically weathered pipe)
The ashen mantle of the phonolithic pipe of the Hohentwiel is well preserved on the western (left hand) side, while on the eastern side it has been removed by the Pleistocene Lake Constance Glacier. Several Miocene remnant volcanoes form characteristic features of the relief in the Hegau area, on the western margin of the Lake Constance Basin.

Hohentwiel near Singen, Hegau, Germany

4.15 Geyser with siliceous sinter cone
Geysers are eruptive springs that mark zones of post-volcanic activity and intermittently eject water and steam. About 200 geysers and numerous hot springs are located in the Yellowstone National Park in the North American Rocky Mountains. Old Faithful is the most famous, ejecting water and steam to a height of 45 metres every 50 to 70 minutes. A characteristic feature of geysers is the formation of siliceous sinter cones (foreground right) around the opening of the geyser pipe. The sinter formations are silicates which precipitate from the hot water as its pressure is reduced.

Upper Basin, Yellowstone National Park, Wyoming, USA

5 Fluvial denudation landforms as a function of climate

Landforms created by fluvial denudation processes are the most common surface forms on the continental areas of the earth and are therefore the first continental surface form to be discussed here. As well as climate, endogenic forces also play a role in fluvial denudation. This is because water flows according to elevation differences, which are primarily determined by endogenic forces. Endogenic control can occasionally even be the dominant factor, leading to the creation of specific structural fluvial denudation forms (Chapter 6). In this chapter, fluvial denudation forms are at first discussed individually and then as a function of climate.

Valleys

Valley cross-sections

Fluvial erosion as a denudational process has already been introduced in Chapter 2. Valleys are formed by water flow in rivers that cut into bedrock. Variations of the form of the valley slopes on either side, and of the valley floor, are the main elements used to typify a valley morphologically. However, a valley cross-section cannot be simply defined as a function of fluvial erosion; it must be related to the processes of denudation and/or water flow on slopes (Chapter 2). An additional role is played by accumulation processes on the valley floor. Three basic types of valley cross-section can be differentiated: the V-shaped valley (Photo 5.1), the V-shaped valley with floodplain (Photo 5.2), and the trough-shaped valley (Photos 5.3, 5.18). Many variations of these basic valley types exist owing to the complex interactions of fluvial erosion and accumulation, of the type and intensity of slope denudation, and of rock layering and lithological types. Extreme types of valley cross-sections are demonstrated in Photo 5.4 (gorge) and in Photo 5.5 (canyon).

Many valley cross-sections are clearly asymmetrical (Photo 5.6) with different slope angles on either side of the valley. Some valleys show step- or ledge-shaped terrace sequences on their slopes. If these are not structurally controlled, they are mostly stream terraces (Photo 5.7) which mark former water levels or valley floors. Fig. 11 shows the sequence of Pliocene and Pleistocene river terraces that occurs along the central part of the Rhine in Germany. Virtually all of the lower terraces are defined as alluvial terraces (gravel terraces) as their alluvial material (gravel) has not yet been completely removed by erosion, unlike most of

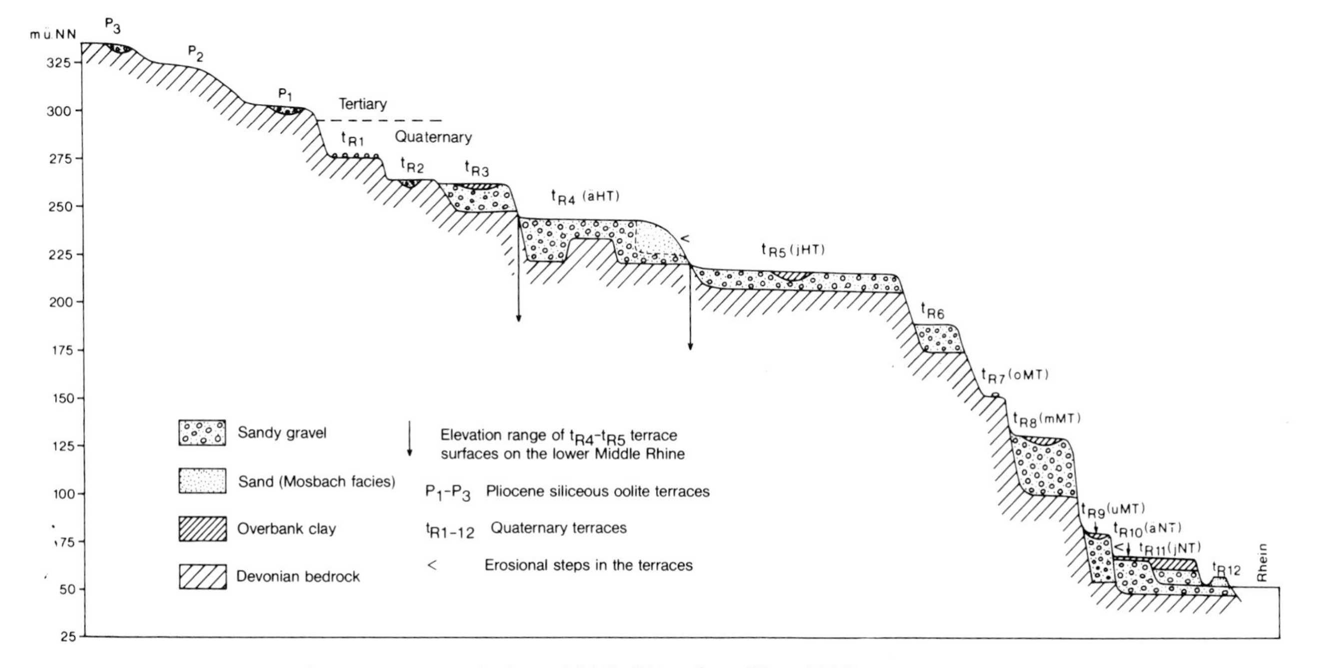

Fig. 11 Schematic section of river terraces on the lower Middle Rhine (from Bibus 1980)
HT - main terrace MT - middle terrace NT - low terrace ä - older j - younger o - upper u - lower

the higher terraces. Deposition of these alluvial terraces and simultaneous lateral corrasion occurred during the Pleistocene cool periods, while erosion dominated during warmer climatic conditions. Terrace sequences such as these along the Rhine therefore indicate frequently alternating, climate-controlled phases of deposition and erosion that can be modified by endogenic processes such as uplift.

Valley profile (longitudinal)

While a valley cross-section can be explained as the result of interactive fluvial and other denudational processes, the (longitudinal) profile of a valley is an expression of the geomorphological effectiveness of flowing water - except for the effects of geologic structure and rock type. Steeper gradients, which are typical features of the upper reaches of rivers, are also areas of deep, effective erosion by rapid water flow movements. Low gradient sections, with reduced flow velocities and smoother flow types, are less affected by erosion even though coarse loads are still transported through these areas. If the gradient is reduced even further, flow velocities can be reduced to a level where accumulation of the coarse load components occurs. Finally, the even lower velocities encountered in the lower reaches of rivers only enable the transportation of suspended and dissolved material (see Chapter 1). If this regular sequence of fluvial process units existed from the source to the mouth of a river, a gradually flattening, concave longitudinal profile (or thalweg) would result. However, this ideal case is never attained owing to the influence of endogenic factors such as geologic structure or rock types which create alternating high and low gradient stretches. In general though, deep erosion in the upper and accumulation in the lower reaches show clearly that fluvial processes generally tend to smooth out the longitudinal profile of a river.

All rivers have sections where abrupt gradient changes or slope breaks - called nick points - appear in the thalweg. They are caused by endogenic factors such as uplift or varying underlying lithologies, and they are often marked by waterfalls (Photos 5.8a, 5.8b), cascades (Photo 5.9), or rapids (Photo 5.10). They indicate points of former headward erosion or valley extension upstream. Headward erosion can result in one stream capturing another stream (Fig. 12), which in turn may lead to fundamental changes in the drainage basins of fluvial systems. Stream capture has been described as stream piracy.

Valley formation in relation to structure

Tectonic control of valley forms is clearly demonstrated by stream incision in mostly V-shaped valleys which penetrate higher elevation areas. The Rhine Valley through the Rhenish Massif between Bingen and Bonn is an excellent example (Photo 5.25). These valleys can be created by either antecedent or epigenetic streams. Antecedent streams traverse a tectonically uplifted block following pre-uplift drainage patterns, and many examples exist in mountain chains (Photo 5.4). Different elevation levels on either side of a mountain range can lead to transverse valley development resulting from headward erosion of streams. Epigenetic (superimposed) streams flow in valleys that were already established on the surface of sediment accumulations which covered older rocks and relief. After erosional incision through the accumulated sediment, a stream can be forced to continue eroding in the same location even when the buried relief is encountered (Photos 5.11a, 5.11b). The structural control of valley development can also be indicated by meandering valleys (Photo 5.12). These winding, deeply incised valleys can result from varying rock resistances in uplifted areas. They can be just as regularly formed as meandering rivers in accumulation

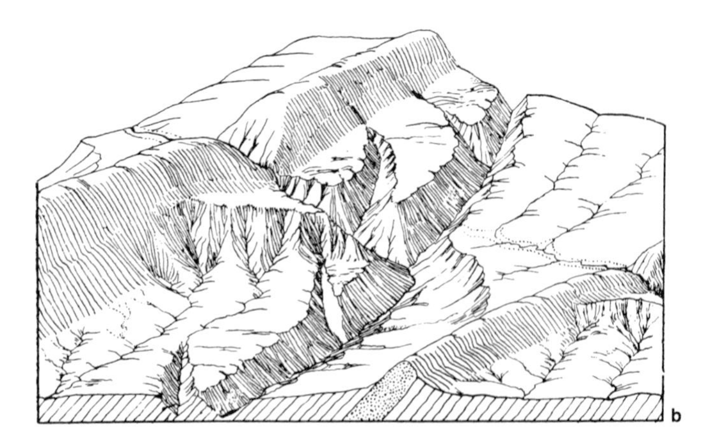

Figs. 12a, b Stream capture (from Davis & Braun 1911)

Fluvial system: **a** before capture, **b** after capture

The stream flowing from left to right in **a** is captured by a river with a lower erosional base due to headward erosion (left side of **b**). Similar captures have resulted in considerable lateral movements of the European water divide between the Danube and the Rhine. Parts of the old drainage basin of the Danube in the Black Forest were captured by headward erosion of tributaries of the Rhine.

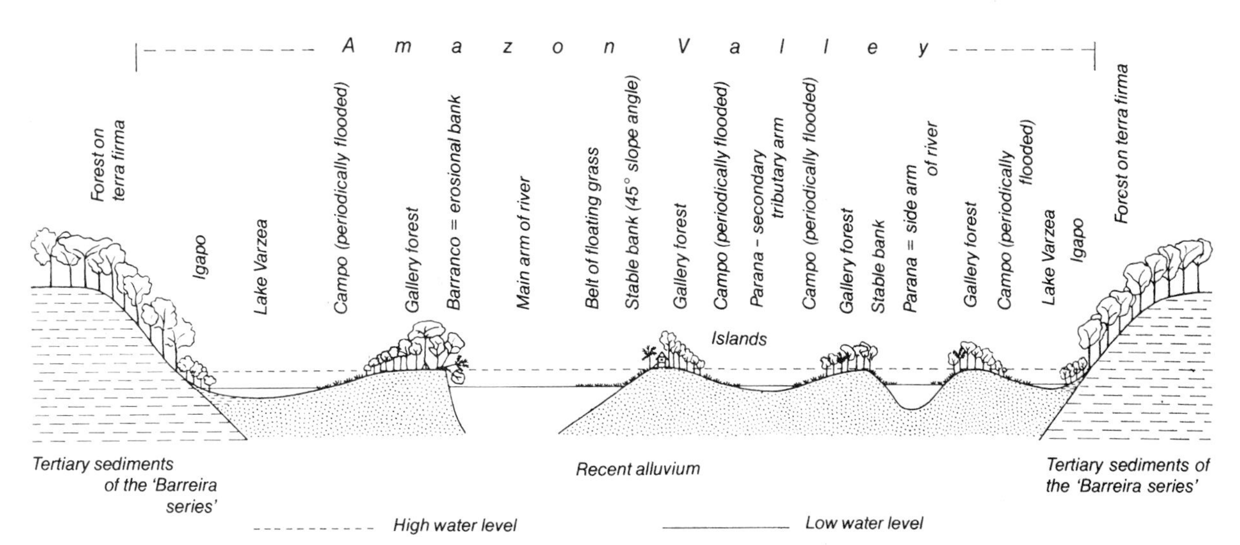

Fig. 13 Levee overflow. Schematic cross-section of the lower Amazon valley, Brazil - vertically exaggerated (from Sioli 1956)

areas, but their path is not controlled by antecedent or epigenetic forms; it evolves during stream incision. Isolated mountains can develop when the ridges between closely-spaced meander loops are removed by erosion (Photos 5.6, 5.13).

Depositional forms

Specific landforms are created not only by fluvial erosion, but also by deposition of coarse to fine-grained material. V-shaped valleys with valley floors covered by alluvium (floodplains; Photo 5.2) have already been mentioned. Coarse and frequently very wide alluvial valley fills or floodplains are often formed by rivers with strong seasonal flow variations (Photo 5.14). Alluvial fans or cones with approximately triangular outlines are a striking fluvial depositional form (Photo 5.15). Natural levées form the banks of sediment-laden rivers on alluvial plains or floodplains (Chapter 3). During floods, the great river systems that are characterized by extensive levée development, such as the Amazon (Fig. 13), Mississippi, Po, etc., deposit coarse sediments on the channel edges or levées and finer material on the adjacent floodplains. This explains the asymmetrical shape of the levées which are steeper on the river side and flatter on the outside, where lakes or swamps can occur. The mean water level of the levéed river can be higher than the adjacent floodplains. Even if the natural levées have been artificially strengthened into dikes, they can be breached by floodwaters, leading to devastating floods and even to lateral channel movements of the river. River meanders (Photo 5.16) are an additional characteristic feature of rivers, especially on floodplains, and deltas are the specific landform that occurs at the mouth of sediment-rich fluvial systems (Photo 5.17).

Plains

The processes on low relief landforms (plains) also enable a differentiation to be made between depositional (accumulation or accretionary) and erosional features.

Accretionary land surfaces

Floodplains are the extensive depositional areas of low relief which are formed by fluvial or marine (Chapter 10) processes. Floodplains of fluvial systems are, however, not exclusively located in the previously-mentioned low-lying areas. They can also be found in

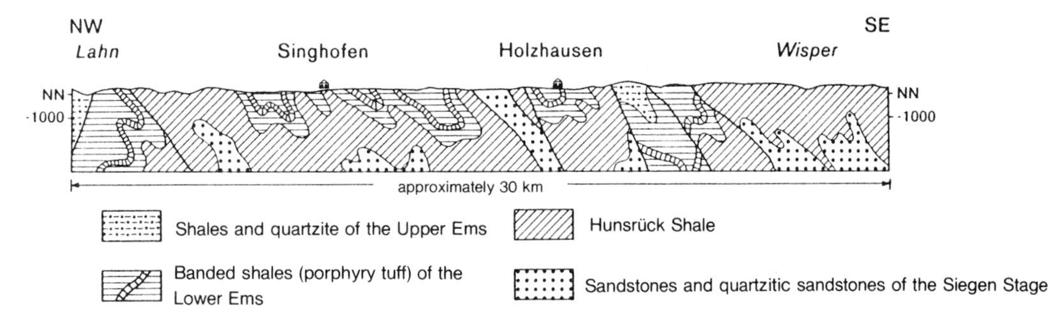

Fig. 14 Erosional (sculptured) surface in the Rhenish Massif, Germany (from Fischer 1986)

A characteristic trait of erosional surfaces is the lack of influence of the underlying rock types and structure. The erosional surfaces in the Central European highlands are fossil or relict surfaces. They were formed by hydrolytic weathering and surface wash under seasonally humid, tropical climatic conditions and were later dissected by rivers and superficially reworked by denudational processes under different climatic conditions.

high altitude basins without drainage outlets, for example in the Great Basin of the North American Rockies or the Altiplano of the South American Cordilleras. Additional, typical accretionary forms are glaciofluvial, coarse or sandy alluvial fans in periglacial areas (Chapter 8), glacis (e.g. lower piedmont slopes), and salt and clay pans in arid areas (Fig. 15).

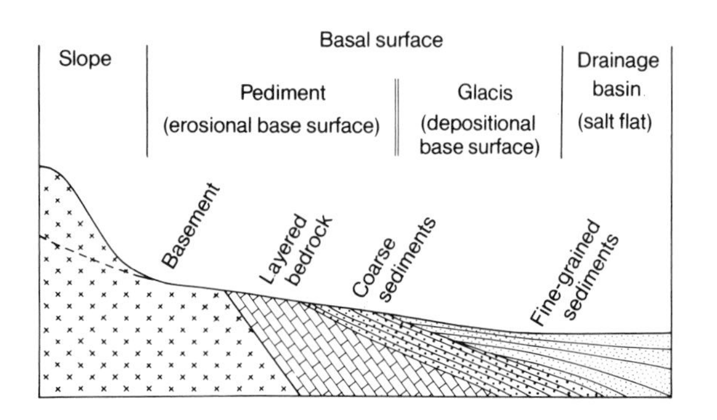

Fig. 15 Schematic section of slope base relief in arid zones (from Ehlers 1980, after Weise 1978)

Erosional land surfaces

A characteristic trait of erosional surfaces is the relative lack of influence of the rock types or their structure. Unlike structurally-controlled surface landforms (Chapter 6), these regional erosional surfaces seem unaffected by the structure of the underlying geology, and are therefore described as sculptured surfaces. They can be subdivided into erosion surfaces (peneplains) and pediments. Erosion surfaces (Fig. 14) are created by the combined effects of rock degradation by hydrolytic weathering and surface runoff (Chapter 2). They are widespread but mostly as relict forms - recently-formed erosion surfaces are limited to the seasonally humid tropics (Photo 5.18). In arid zones, pediments (Photo 5.19) represent the erosional zone at the base of mountain fronts formed in bedrock (Fig. 15). Further downslope they merge into gently-inclined glacis slopes - the closest depositional zone to the erosion surface.

Climate control

This overview of the basic types of fluvial denudation landforms indicates clearly that their formation results from a combination of fluvial and other geomorphological processes. Global zones with similar process combinations can be recognized, and this shows that geomorphological processes are fundamentally climate-controlled: zones with similar process combinations can be closely correlated with climatic zones suggested originally by Köppen. Table 3 contains a simplified schematic comparison of these process zones with their characteristic landforms, as well as their climatic characteristics. The following text is a brief summary, from the poles to the equator, of the dominantly fluvial denudation landforms and their differentiation as a function of climate.

Subpolar (I) and cool-temperate climates (II). The Roman numerals in brackets refer to the climate zones as conventionally described in the Köppen system. The polar climatic zones, which feature glacial geomorphological processes (Chapter 8), are followed by the subpolar and cool-temperate climatic zones. Seasonal freeze-thaw cycles are a characteristic feature of both climatic zones. Weathering and erosional processes are very intensive. Summer thawing of the surface soil layers on top of the permanently frozen (permafrost) ground, together with frequent diurnal freeze-thaw cycles, can induce active frost processes. Microsolifluction is a material sorting process in inhomogeneous debris deposits that forms frost-patterned soils (Photos 2.16, 5.20). Gravity induces intensive soil flow (gelisolifluction; Photo 2.16), even on very low-angle slopes. Diurnal freeze-thaw cycles with intensified frost weathering during the warmer months (Photo 2.2), lead to the accumulation of extensive debris cones in mountainous terrains (Photo 2.11), and - together with solifluction - to characteristic smooth slope forms (Photo 5.21). Surficial water flow is restricted (80–90 per cent) to the brief snow melting periods, during which subsurface soils are, however, still frozen and impermeable. Surface wash processes on slopes and fluvial processes are therefore quite intensive for short periods of time. Slope serration (Photo 6.2) and broad alluvial valley fills (Photo 5.22) are the results of active fluvial processes. Many typical landforms of the subpolar and cold-temperate climatic zones can also be found in the corresponding temperature zones at high altitudes in cool-temperate, subtropical and tropical zones.

In the *humid areas of the cool-temperate climatic zones (IIIh)*, geomorphological activity is relatively slight. As previously described, denudation processes are rarely obvious, with the exception of slow gravitational mass movements (creep) and isolated slide and glide occurrences (Photos 2.19, 2.20). Effective local denudation processes are mostly the result of anthropogenic effects. Erosion in this climatic zone is mostly initiated by chemical weathering, especially by solution processes, as indicated by fluvial loads that are predominantly transported in solution (Chapter 1). The main reason for the absence of spectacular erosional forms is the comparative lack of extreme climatic conditions, such as long or intensive rainfall periods. It is indicative that intensive, spontaneous erosional processes only occur locally under these extreme weather conditions, and only if particular local lithological and structural conditions exist (e.g., block slide with rock fall; Photo 2.18). The present landforms in the humid areas of the cool-temperate climatic zones are therefore mostly predetermined by older, relict landforms. Regional erosional surfaces, which are now cut by fluvial systems and superficially reworked by denudational processes, were inherited from earlier (Tertiary) relief generations (Fig. 14, Photos 5.23, 5.24). Younger Pleistocene generations are indicated by V-shaped valleys with or without floodplains (Photos 5.1, 5.2, 5.25), and river terraces, as well as by typical depressions (Photo 5.26) created by Pleistocene glacial processes, and leading to characteristic micro- and mesoforms. A transitional area between the relief of an erosion surface and a flat-lying relief is shown in Photo 5.27.

In the *arid areas of the cool-temperate climatic zones (IIIa)*, as well as in *warm-temperate subtropic areas* with humid winters and dry summers *(IVh)*, which are often described as Mediterranean, both physical and chemical processes play a role in shaping the land. Both climatic zones are characterized by periodic heavy rainfall as well as by anthropogenic removal of vegetation (especially in the Mediterranean type) resulting in intensive slope erosion. Fluvial processes are active during short periods of high intensity rainfall, in winter in the Mediterranean, and during summer in the arid areas of the cool-temperate zones. Common landforms are debris cones, pediments (Fig. 15), and serrated slopes (Photo 5.28), while valley floors are covered by extensive alluvium (Photos 5.30, 5.31).

The *arid areas of the warm-temperate subtropics (IVa)* and the *tropics (Va)* are formed by complex process combinations. Strong insolation weathering (Photo 2.1) dominates, together with frost weathering in the subtropics. Slopes are covered by debris fans, and plateaux by debris and boulder mantles (Hamada; Photos 5.32, 5.33) created by water and wind corrasion. Duricrust formation is common (Photos 2.7, 2.8, 5.35). Lack of vegetation, combined with occasional heavy precipitation, leads to effective slope (Photo 2.25) and surface wash (Photo 2.26) processes, which occur periodically in semi-arid areas and episodically in completely arid zones. A typical relief sequence is formed at the base of mountains consisting of pediment, glacis, and salt or clay flats (Fig. 15, Photos 5.19, 5.34, 5.35, 5.36, 5.37, 5.38). Rivers in arid climates are either perennial, allogenic or exotic rivers such as the Nile or Indus, that have external sources and manage to traverse the arid zone successfully, or endoreic-type systems that originate in a humid area and end in an arid inland area, such as the Hilmend in Afghanistan. Most valleys in arid zones are, however, dry valleys (Wadis; Photo 5.39) with periodic or, in completely arid areas, only episodic water flow in limited sections of the braided river. They are relict forms originally created in less arid conditions during the Pleistocene. Most of the extensive stony desert areas of the Sahara (Serir; Photo 5.41) were also formed under less arid conditions than presently typical for the area. Inselbergs (Photo 5.40), which were created under seasonally humid tropical climatic conditions, are typical relict forms of older relief generations. Tors (Photo 5.42) have polygenetic origins as several generations of processes contribute to their formation. Recent geomorphological activity in arid zones is particularly strongly affected by aeolian processes which are discussed in more detail in Chapter 9.

The *warm-temperate subtropics with humid summers (IVh)*, on the equatorial side of the arid zones, are restricted to the eastern sides of the continents and are affected by both chemical and physical weathering. This is especially valid in the *seasonally humid tropics with low humidity (Vh)*, i.e. with long dry periods. Periodic strong precipitation events in the extensive areas of the seasonally humid tropics trigger intensive surface wash. In mountainous terrains, which are quite common in this climatic zone, slope wash is highly active and erosion is relatively rapid. Erosion surfaces (peneplains) with broad valleys (Photo 5.18) and inselbergs (Photos 5.41, 5.43) dominate in areas of flat-lying relief. Elsewhere, V-shaped valleys with floodplains (Photos 5.45, 5.46), periodic water flow, and serrated slopes (Photo 2.23) are the most common forms. Intramontane plains in mountainous areas must be regarded as initial stages of erosion surface (peneplain) formation. They are surfaces that occur along sections of large rivers where tabular landforms, a result of the combined effects of hydrolytic weathering and surface runoff, extend to elevated land surfaces.

Humidity levels increase considerably towards the equator in the warm-temperate subtropical zone with humid summers and in the seasonally humid tropics with low humidity. The next climatic zones are the *perenially humid subtropics* and the *seasonally humid tropics with high humidities*, both of which have shorter dry seasons. They represent an intermediate zone of morphodynamic process combinations: the area is affected by processes that are typical of both the adjacent zones – toward the poles as well as toward the equator (Table 2). Geomorphological processes activated by perenially high humidities and temperatures are typical for the *zone of tropical rainfall (Vh)*, and become increasingly important closer to the equator. These processes include intensive chemical weathering, intensive fluvial activity and strong gravitational mass movements. Tropical relief is therefore characterized by aggressive valley incision, with V-shaped valleys having floodplains in flat-lying areas and steeply sloped V-shaped valleys (Photo 5.47) in mountainous areas. Gravitational mass movements do not only occur as slides, slips, or block glides (Photo 5.48), but also as sub-forest solifluction. This type of subterranean mass movement beneath forests is favoured by the development (largely by hydrolysis and oxidation) of thick, moist regolith (Chapter 2). Termite mounds are striking organogenic microforms in tropical savanna and forest areas (Photo 5.49).

5.1 V-shaped valley

The V-shaped valley is the first of the three basic cross-sectional types used here. They are deeply incised valleys with symmetrical slopes on either side. V-shaped valleys indicate strong incision and slop erosion resulting from tectonic uplift. The example shown is a river meander in a V-shaped valley (flow direction of the river on the side towards the viewer, and in the opposite direction on the left).

Loop of the Saar River near Mettlach, Saarland, Germany

5.2 V-shaped valley with floodplain

The second basic type of valley cross-section can be differentiated from the V-shaped valley by the presence of a valley floor; in its most extreme form it can be described as 'box-shaped'. Two processes are usually responsible for its formation: first, the valley floor can be created by lateral corrasion of the bedrock and therefore represents a bedrock base which is covered by a thin layer of alluvium. Secondly, in highland areas under humid, cool-temperate climatic conditions, V-shaped valleys with floodplains were originally true V-shaped valleys which were then partially filled under Pleistocene glacial conditions. In the example shown here, the river has removed several metres of alluvium in the channels since the end of the last glacial period.

Sauer Valley near Goebelsmuehle in Oesling, Luxembourg

5.3 Trough-shaped valley

Different origins can account for the same general morphology of this valley type. Trough-shaped valleys in the seasonally humid tropics (see Photo 5.18) are created by the combined effects of linear erosion and runoff processes on slopes. On the other hand, trough-shaped valleys associated with the upper reaches of fluvial systems in humid areas of cool-temperate climates are mostly derived from V-shaped valleys. Concave slope forms are explained by increasing thickness of slope debris towards the base of the slope. The debris filled the valleys during the Pleistocene glacial periods as a result of gelisolifluction processes. Trough-shaped valleys can also result from glacial processes (Photo 8.12).

Incised meander of the Neckar River near Neckargerach, Kleiner Odenwald, Germany

5.4 Gorge

Gorges are extreme types of V-shaped valleys. Characteristic features are vertical valley walls, great depth, narrow width, and the absence of a valley floor. Gorges indicate strong erosional forces that tend to follow weak structural or lithological zones. They are mostly relatively young features as the angles of the valley slopes have not yet been lowered.

Rio Santa, Cordillera Negra, Peru

5.5 Canyon

Canyons are also variations of V-shaped valleys. Valley floors have not been created by strong erosional activities, and the valley walls are stepped, reflecting alternating rock resistances in the eroded, horizontally-bedded sequences. The most famous example is the Grand Canyon of the Colorado River. The river has cut down to a depth of 1800 metres, 1500 metres of which were through horizontally-bedded sediments, followed by 300 metres of underlying granites. Maximum horizontal width at the upper edge is 25 km.

Grand Canyon of the Colorado River, Arizona, USA

5.6 Asymmetric valley

Different slope angles on opposing valley sides can be particularly impressive in meanders. The outer sides of meanders are undercut by erosion leading to the formation of steep bluffs, while the inside of the meanders have shallow-angled slopes (slip-off slopes) or point bar deposits. In the foreground of the example, the inner slip-off slope flattens to the right, while a steep bluff can be seen in the left background. Additional reasons for asymmetric valley formation include the degree and direction of exposure to moisture (rainfall), desiccation (sunlight), and – in periglacial areas – thaw processes leading to variations in the strength of slope denudation processes.

Our Valley near Bivels in the Oesling area, Luxembourg

5.7 River terrace

River terraces are remnants of older valley floors that have been destroyed by erosion and denudation (Fig. 11). The example shows a cut through an old valley floor which is very wide where two valleys join in the foreground. In the background the former valley floor has been completely removed in several sections by slope erosion. Occasionally, however, it is still visible as a debris-covered ledge on the far slope.

Tributary valley of the Rio Grande, Sierra de Chani, Jujuy Province, Argentina

5.8a Waterfall (river)

Waterfalls mark the most impressive steps in the longitudinal profiles of valleys, and are mostly caused by resistance variations of the underlying bedrock. The Niagara Falls between Lake Erie and Lake Ontario are two 55-metre-high falls separated by an island, and were created by highly resistant Silurian limestones and dolomites. Below the falls, the Niagara River flows in an 11-km-long canyon that was formed by headward erosion since the last Pleistocene glacial period. The headward incision rate used to be 0·6 m per year for the American Falls and 1 m per year for the Canadian Horseshoe Falls. The high rates were caused by continual undermining and collapse of the limestones and dolomites which overly less resistant shales. The cementation of the base of the falls during power station construction in 1969 reduced the incision rates to 2 and 6 cm per year for the American and Canadian falls respectively.

American Falls, Niagara River, New York State, USA

5.9 Cascades

Cascades are a sequence of small waterfalls (waterfall steps). This form of fluvial activity is also formed by resistance differences in the underlying bedrocks.

White Nile, below Lake Victoria, Uganda

5.8b Waterfall (stream)

In Minneapolis the Mississippi forms the 12-metre-high St Anthony Falls which led to the founding of the city. Just to the south of the city centre, Minnehaha Creek has formed a waterfall that is also controlled by an outcrop of the 9 m thick, resistant, horizontal Silurian Platteville limestone. The St Anthony Falls – similar to the Niagara Falls – have been affected by considerable headward erosion since the end of the last Pleistocene glacial period. Such erosion has not happened to the Minnehaha waterfall because the erosive force of the small 12-km-long creek was insufficient. Some undercutting is, however, occurring at the base of the falls, leading to slight headward erosion.

Minnehaha Falls, Minneapolis, Minnesota, USA

5.10 Rapids
Rapids or cataracts mark changes in gradient of the long profile, whereby rivers traverse large elevation differences over a short distance. Water movements are characterized by extreme turbulence, and water depths vary greatly.

Rivière Montmorency, Laurentides National Park, Quebec, Canada

5.11a River valley (antecedent or epigenetic?)
The Wind River cuts through the Owl Creek Range in a V-shaped valley. The range is a dome-shaped feature with a core of Precambrian magmatites and metamorphics and the overlying Paleozoic and Mesozoic layers form ridges that face the centre of the dome. The photograph shows the Wind River emerging from its incised valley on to an alluvial plain. The Wind River valley can be explained by either antecedent or epigenetic (superimposed) valley formation. Either the dome of the Owl Creek Range is younger than the river course (antecedent), or the original valley was located on sediment accumulations that have since been removed (epigenetic). In the latter case, the Wind River will have first cut into the accumulations and then into the underlying bedrock. An epigenetic origin is suspected for the incision of the Wind River Canyon.

Wind River south of Thermopolis, Wyoming, USA

5.11b Epigenetic river valley
Examples of obviously epigenetic valley formation can be seen on the southern edge of the Fränkischer Alb and the Bayerischer Wald in southern Germany, where the Danube flows in incised V-shaped valleys along certain stretches of the river. These valley stretches were originally located on the surface of an alluvial cover, of which some small sections still remain. As a result of continuing uplift, the river was forced to cut into solid bedrock.

Danube Valley near Weltenburg, Germany

5.12 Meandering valleys

Valley meanders are formed by the erosional incision of river meanders into the bedrock following tectonic uplift. The valley itself has a meandering form. The meanders are mostly of the same order of size, and adaptations to different rock resistances can be observed.

Ceyhan Valley near Karatepe, northeast of Adana, Turkey

5.13 Meander core

Continued erosion of the ridge between closely-spaced valley meander loops finally enables the river to cut through the intervening ridge. The truncated loop then surrounds an isolated meander core within the abandoned meander. The meander core in the centre of the photograph is encircled by a meander of the Vis river. After the ridge was breached (foreground), the valley floor in the former meander dried up.

Cirque de Navacelles, Causse du Larzac, Massif Central, France

5.14 Alluvial floodplain

Alluvial floodplains are formed by rivers with strong seasonal flow variations. During low water periods, most of the alluvial plain is dry, but high water levels can cover the entire width of the floodplain. Coarse material is transported in pulses during flood periods, i.e. it is transported short distances, temporarily deposited, and then picked up during the next flood. The individual channels into which the river withdraws during low water periods often change their course across the plain, resulting in a so-called anastomosing or braided streams. The microrelief on the surface of alluvial floodplains changes constantly due to alternating water flow, transport, and deposition – if unchecked by human intervention (dam, channel construction).

Rio Caldera, Valle de Lerma, Salta Province, Argentina

5.15 Alluvial fan, alluvial cone

Alluvial fans or cones with triangular outlines are formed when streams with coarse bedloads encounter sharply reduced gradients, for example, when they leave mountain valleys and emerge onto a plain. The coarse components accumulate in fan-shaped deposits. Low, relatively flat deposits are described as alluvial fans, while fans with steep slopes form alluvial cones. Channels frequently cut into the surfaces of fans and move laterally across them, as on alluvial floodplains. This occurs when deposited coarse material blocks and deflects the intermittent water flow.

Western flank of the Sangre de Cristo Range, south of Salida, Colorado, USA

5.17 Delta (right)

Deltas are sediment accumulations that form at the mouths of sediment-rich rivers when they reach a large water body such as the ocean or a lake. The triangular shape of deltas is created by the separation of the main river channel into individual channels that repeatedly change their course. Sediments accumulate on levées along channels or as overbank deposits between channels. Delta growth is most pronounced on flat coastlines that have low tidal ranges. Deltas form on a complex system of superimposed sedimentary units that were deposited by earlier river channels with different courses. Delta thicknesses can vary widely. Sediment thicknesses in the Mississippi delta can reach 1000 m, as the coastal zone is subsiding beneath the sediment load. Up to 400 million tons of material are deposited in the Mississippi delta every year, and the most rapidly growing arm is presently moving seawards at a rate of nearly 100 m per year.

Mississippi delta, Louisiana, USA (photograph: NASA, E-1177-16023; from Kronberg 1985)

5.16 Meandering river

Meander curves are formed by the dynamics of linear free-flowing water, and meandering rivers occur in fluvial accumulation areas with very low gradients. The radius of the meander curves is controlled by the size of the river, i.e. by its water volume. Meander channels often move in a meander belt, the width of which is determined by the radius of the meandering curves. Accumulation occurs on point-bars on the inner sides of meanders, while erosion undercuts the outer flanks of the curves. Destruction of the meander neck between two closely-spaced loops leads to the formation of ox bow lakes and meander cores (see Photo 5.13).

Yukon River in the Yukon Flats floodplain near Fort Yukon, Alaska

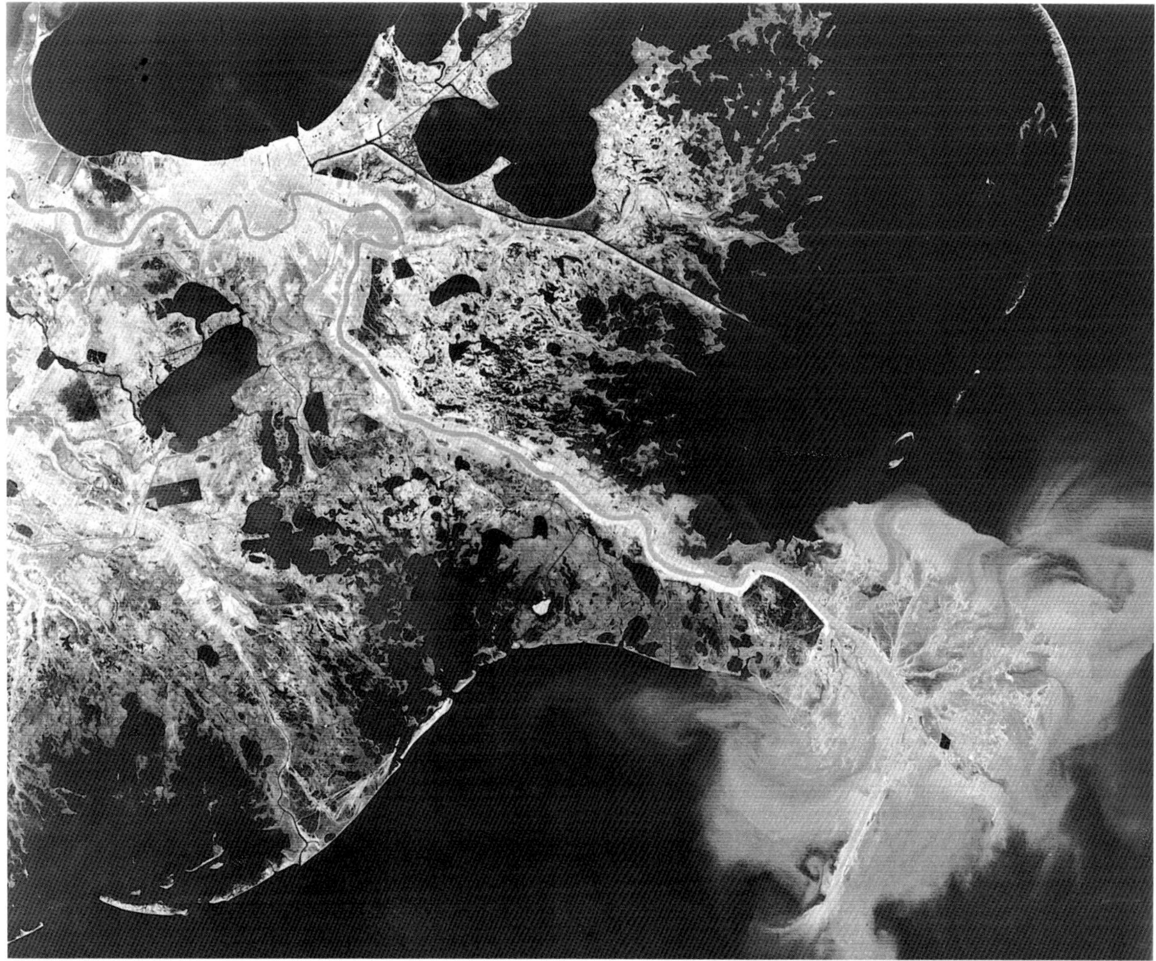

5.18 Erosion surface with trough-shaped valley
Erosion surfaces (peneplains) exist at various elevations and are created by the combined effects of hydrolytic weathering and surface runoff. They are not affected by the underlying rocks and form a characteristic low relief with shallow trough-shaped valleys. Periodic surface wash erodes the low angle slopes. Recent erosion surfaces are limited to the seasonally humid tropics.

Brazilian Highlands (Planalto) near Ribeirão Preto, São Paulo State, Brazil

5.19 Pediment
Pediments are erosional surfaces with slightly steeper slopes (3–8°) than peneplains. They represent the uppermost units of slope base relief (Fig. 15), and are formed at the base of mountains in arid zones by a combination of slope degradation (brought about by physical weathering and slope runoff), surface wash, and lateral erosion by streams emerging from the mountains. All of these processes are only periodic or episodic. The example shown has developed in granite and is covered by a thin debris mantle.

San Bernadino Mountains, Southern California, USA (photograph by H. Mensching, Göttingen, Germany)

5.20 Patterned ground
Patterned or polygonal ground mostly occurs in zones with seasonal and frequent diurnal freeze-thaw cycles. Repeated freezing and thawing of loose surface materials induces movements on level surfaces (cryoturbation) which lead to material sorting. This type of autochthonous mass movement is also described as microsolifluction, and it creates circular or polygonal patterns marked by coarse debris at the edges and increasingly finer material towards the centre of the polygon. It is effective down to the depth influenced by freeze-thaw cycles. The diameter of the polygons is controlled by the penetration depth of the freeze-thaw cycles and can range from several centimetres to more than 1 metre (see hammer as scale in photograph). On very shallow slopes (2°), the polygons are stretched into parallel stone lines separated by fine-grained zones (Photo 2.16).

near Ny Ålesund, Spitsbergen

5.21 Smooth slope

Smooth slope forms are created by strong frost weathering together with slow gravitational mass movements (soil creep), and occur in subpolar and cool-temperate climatic zones, as well as in thermally equivalent high-altitude zones. Characteristic features are smooth and relatively steep slope profiles. Smooth slopes occur when bedrock has disintegrated by frost weathering into small homogeneous fragments.

near the Abra de Pives Pass (4170 m), Jujuy Province, Argentina

5.22 V-shaped valley with alluvial floodplain

V-shaped valleys with floodplains that occur as relict forms (Photo 5.2) in highland areas in cool-temperate climatic zones are the characteristic valley forms of subpolar and cold-temperate climates. High water flow rates, even though they are limited to the short snow melt periods, mean that the rivers are strong erosional and transport agents. Wide, gravel-covered floodplains are formed (see also Photo 5.31). Effective lateral erosion is indicated by the steep boundaries of the valley floor in the tributary which enters the main valley on an alluvial fan.

Advent Valley and tributary valley, Spitsbergen

5.23 Remnant erosion surface

Erosion surfaces (peneplains) in the humid areas of the cool-temperate climatic zone are inherited from earlier relief generations and have been superficially reworked. They were formed under seasonally humid tropical climatic conditions during the Tertiary. An important feature is their structural independence (Fig. 14, Photo 5.18). Reworking was performed by erosional and denudational processes during the Pleistocene: erosional incision during the Pleistocene warm periods, and denudation during the Pleistocene glacial periods. The original relief of the erosion surface can be theoretically reconstructed by 'removing' the deeply incised valleys. Denudational erosional processes (gelisolifluction) have almost completely removed the hydrolytic-weathered material – a typical feature of present erosion surfaces – during the Pleistocene glacial periods. Hollow forms (Photo 5.26) created during the Pleistocene glacial periods are common features of the present surface relief.

Oesling highlands west of Vianden, Luxembourg

5.24 Incised remnant erosion surface
Deep incision of old erosion surfaces – a common feature in areas of strong tectonic uplift – often only leaves isolated remnants of the old land surfaces. In Central Europe, this process affects the marginal areas of the Upper Rhine Graben, for example the tilted block of the uplifted southern part of the Central Black Forest (Fig. 8).

View from the Belchen towards the Feldberg, Central Black Forest, Germany

5.25 V-shaped valley with river terraces
Fluvial erosion during the Pleistocene warm periods has frequently destroyed older low-relief surfaces. In the example shown, the Rhenish Massif has been dissected by the Rhine, which has formed a V-shaped valley. The river is a powerful erosive agent and quartzite ridges used to form rapids at this location, but most of them have been removed from the river bed in order to reduce the danger to shipping. Alternating erosion and accumulation processes are indicated by terrace formation (Fig. 11) at various elevations (on the left side in the centre of the photograph, on the right side in the background).

Upper part of the Central Rhine Valley, downstream view from the Loreley Rock, Germany

5.26 Dell
Dells associated with Pleistocene glacial periods are particularly striking geomorphodynamic forms in the humid cool-temperate climatic zones. They occur as micro- or meso-forms on remnant erosion surfaces as well as on structurally-controlled terraced surfaces. They are mostly flat, extended features with symmetrical, even slopes, and broad, trough-shaped cross-sections unmarked by stream channels. Larger surfaces can be covered by branched systems of dells. They merge into a main dry valley and then into a valley marked by the appearance of a stream channel. The dells were originally formed as meltwater channels that were repeatedly filled by laterally intruding solifluction masses. They therefore document two strong erosional processes that were active under Pleistocene glacial conditions: first, surficial gelisolifluction in thawed ground over frozen, impermeable subsoil, and secondly, linear flow of meltwaters which transported solifluction material into the valleys.

Terraced surface of the Muschelkalk southwest of Oberndorf/Neckar, Germany

5.27 Transition between structurally independent and dependent forms
A transitional zone between the structurally controlled (sculptured) forms of peneplains and the structurally independent forms of flat-lying relief (see Chapter 6) is apparent in the northern Black Forest. The granites of the southern Black Forest are covered in the northern areas by Bunter Sandstone (Lower Triassic) sediments. The Braunberg, a Bunter Sandstone butte, is shown in the centre of the photograph, while in the background the flat-lying Bunter Sandstone can still be seen over the pre-Permian erosion surface of the underlying rocks. Pre-Permian remnants can also be seen in the gently dipping slopes at the base of the butte. The Rench valley in the foreground has cut deeply into the erosion surface and its tributary streams have nearly completely destroyed the old erosion surface, which was uncovered by the receding Bunter Sandstone. Only the uppermost reaches of the streams in front of the Braunberg attain the level of the old erosion surface.
Braunberg near Bad Peterstal, Black Forest, Germany

5.28 Badlands
Areas with extreme slope serration are called badlands in North America. They are a characteristic feature of all semi-arid areas of the cool-temperate climatic zone and are formed by rill wash in relatively soft rocks during heavy precipitation. The removal of surface material by erosional-denudational forces is so rapid that vegetation cover cannot evolve. Anthropogenic destruction of vegetation, or even the temporary removal of dead vegetation, can trigger their formation. In the example shown, flat-lying Oligocene clays and marls have been serrated by fluvial processes and slope wash.

Badlands National Monument, east of Rapid City, South Dakota, USA

5.29 Valley in steppe zone
A characteristic feature of valleys in semi-arid steppe zones are dry stream beds with only periodic or episodic water flow which can vary widely in strength. In the steppe plateaux of northern Iraq, which are covered by loess-type soil, rivers only transport material in solution or in suspension. Occasional lateral corrasion, indicated by the steep banks of the shallow valley, results from the concentrated effects of flood waters after heavy rainfall. The construction of a concrete floodway or ford is sufficient to enable road transport throughout most of the year.
Steppe plateau west of Mosul, northern Iraq

5.30 Mountain valley

The dry river bed in the V-shaped valley transports large amounts of coarse material during the episodic winter and spring floods. Corrasion not only occurs on the river bed but also laterally, as indicated by the steep river banks. The example shown is on the southern edge of the Djebel Sinjar, an anticlinal structure that appears as an isolated highland area in the northern Iraqi steppe plateau, and which is the source area of the stream in the example. Annual mean rainfall in the Djebel Sinjar area is just over 400 mm, with more than 90 per cent during the months November to April.

Djebel Sinjar, northern Iraq

5.31 Torrent

The Italian word *torrente* is not just used for a rapid stream of water, but also as a geomorphological term for a typical valley type in Mediterranean climatic zones (moist winter/dry summer areas in warm-temperate climates). The upper reach is a V-shaped valley with a gravel-filled bed (Photo 5.30), while the downstream sections are covered by broad pebbly beds, owing to periodic strong water flow and concomitant high transport energy (Photo 5.14). Both stream bed and lateral erosion can be quite substantial.

Ebro Basin, south of Lérida, Spain

5.32 Debris Hamada

Rock debris created by physical weathering (insolation and frost weathering) and often coated by manganese crusts, covers extensive areas of desert surfaces called hamada. The hamada shown here is covered by basalt debris. The fine-grained material created by erosion is usually swept away by wind action, but can also be removed by water wash.

Syrian-Iraqi desert plateau southeast of Djebel Drus, Jordan

5.33 Rock Hamada

The surface of platy, disintegrated sandstones is covered by isolated flat, sharp-edged rock fragments that often show signs of wind corrasion. Processes of formation are identical to those described for Photo 5.32.

Terraced surface of the Jilh Sandstone, east of Buraidah, Saudi Arabia

5.34 Glacis

Debris-covered, gently inclined (to 3°) glacis slopes at the base of mountain ranges lie beyond pediment surfaces (Fig. 15, Photo 5.19). While pediments are mostly relatively narrow features (background), glacis surfaces cover broad expanses marked by channels of linear, episodically-flowing streams (foreground) and lead into salt or clay flats. Repeated episodic surface wash, accumulation, and erosion has left the glacis covered by coarse material. Fine-grained material is deposited in the salt and clay pans unless it is completely removed by wind erosion. The thickness of the accumulated debris increases with distance away from the base of the mountain range.

West of Kandahar, Afghanistan

5.35 Glacis terraces

Glacis surfaces are frequently cut into individual terraces by erosion. These step-shaped systems of glacis terraces are often caused by climate-induced morphodynamics. For example, during the Pleistocene, extremely arid phases alternated with less arid phases. Just behind the recent erosive cut, the lowest terrace of the glacis terrace sequence can be seen (foreground) with the remnants of a higher glacis terrace behind it. This glacis terrace system has been deposited at the base of an erosional scarp in the central Sahara. All rock fragments on the surface of the glacis terraces are covered by a manganese duricrust formed by capillary transport of Si- and Al-rich solutions.

Terraced base of the Adrar Acacus near Ghat, central Sahara, Libya

5.36 Dry lake (playa)

Undrained and often tectonically-controlled shallow areas, which are frequently described by local terms (playa, sebkha, salina), represent the final stage of characteristic mesorelief features in arid environments and follow pediment and glacis areas (Fig. 15). These dry lakebeds (salt or clay flats) are episodically or periodically filled by shallow lakes and thick layers of fine-grained material can result. Desiccation of the lakes typically produces mud cracks (see Photo 5.37) and salt encrustations. Salt crusts are common in semi-arid areas and salt swamps or lakes are formed during flood periods (Photo 5.38). The example shown is located at an altitude of 3500 m on the Andean Altiplano.

Salinas Grandes, Jujuy Province, Argentina

5.37 Wadi (oued)

Wadis are dry valleys in arid zones. They are mostly box-shaped valleys with broad gravel floors and steep flanks. Water flow is only episodic or periodic, but is often of great strength, with high levels of material transport and bed and lateral corrasion. Gradient inversions in the stream channels as a result of fluvial accumulation or windblown sand are overcome by the pulsed flow events. The wadis and their tributary valleys are relict features that were created under more humid climatic conditions. This is also indicated by the fact that in completely arid areas, episodic water flow and reforming of the river beds only occurs along subsections of the wadis.

Oued El Abiod, southern edge of the Sahara-Atlas, Algeria

5.38 Salt lake

Salt lakes (local names: kewir, salina, shott), which represent one of the forms of dry lakes (Photo 5.36), often feature large polygons (diameter 1–3 m; for scale see sunglasses in the photograph). These forms are created by the combined effects of moisturization, solar radiation, temperature, and wind. Evaporation first forms a whitish salt crust on top of the saline mud. Further desiccation leads to the formation of desiccation cracks in the salt crust, which are marked by upturned edges. These cracks can reach the deeper clay layers, and thermally-induced upward movements of saline mud commence. Intensive evaporation leads to constant salt crystallization below the cracks that pushes up the polygon edges.

Shott Melrhir, Algeria

5.39 Desiccation cracks in clay pan

Dry lakes, or salt or clay pans, are the final stage of the typical morphological sequence in arid zones (Fig. 15). They are not restricted to subtropical and tropical arid zones, but can also occur in arid areas of the cool-temperate zone. After the periodic water coverage has evaporated, a polygonal network of desiccation cracks forms on the surface of fine-grained material that has been washed into the dry lakes. Crack widths are approximately proportional to polygon size.

Edge of the Tüzgölü, Central Anatolia, Turkey

5.40 Serir

A serir is a gravel- or pebble-covered desert plain. In the glacis zone (Fig. 15), accumulated material is rounded during transport by water. However, such movements mostly occurred during earlier, less arid periods. This serir has been incised by erosion. On some serir plains, the gravel and pebble material is often an autochtonous, resistant remnant of congolmeratic sandstones. The fine-grained material has been removed by water or wind.

Foothills of the Djebel Ben Ghnema, Central Sahara, Libya

5.41 Inselberg

Inselbergs formed under seasonally humid tropical conditions, are strongly reworked relict forms. In the arid interior areas of Australia, Ayers Rock is probably the most famous inselberg as well as being the largest monolith in the world. About 30 km away, the Olgas are a group of dome-shaped inselbergs with a maximum elevation of 1040 metres. This is approximately 500 metres above the surrounding fossil peneplain. While Ayers Rock consists of Cambrian sandstones, the Olgas are formed by resistant conglomerates with dip angles of 20°. Vertical fractures allowed the original monolithic inselberg to be subdivided into individual mountains under changing morphodynamic processes related to repeated climatic changes. The Olgas represent, like all inselbergs, weakly fractured erosional remnants. Their emergence from the surrounding peneplains is explained by their relatively high erosional resistance.

The Olgas, Northern Territory, Australia

5.42 Tor

Tors are polygenetic forms composed of rounded blocks of massive rock types such as granite. Tors can form in a number of ways. In this case, the blocks were the 'floating' remnants unaffected by chemical erosion, in a ferrosiallitic weathering residue which was formed under seasonally humid tropical climatic conditions (Chapter 2, Photo 2.4). Removal of the weathered residue left the blocks as isolated rocks or as rock piles. The whitish, decomposed material, on the hills in the background, is an old erosional feature of the granitic basement material, covered by terrace-forming Paleozoic sandstones. The receding terraces exposed the weathered residue and enabled its removal. The residual granite blocks are now exposed to strong physical and physico-chemical weathering forces.

Asir Highlands, north of Najran, Saudi Arabia

5.43 Erosion surface (peneplain) with inselberg
Erosion surfaces (or peneplains) are the dominant macroform in the seasonally humid tropics. The only structural features of extensive peneplains are very broad, trough-shaped valleys (Photo 5.18). The most striking mesoforms on peneplains are, however, inselbergs, which rise isolated or in groups (Photo 5.41) above the low relief of the surrounding country. They are erosional remnants of peneplains formed by concomitant hydrolytic weathering and surface wash. Steep slopes and sharp basal angles indicate active slope recession.

near Ferkessedougou, Ivory Coast

5.44 Balanced rock
If the hydrolytically-formed residue in peneplain areas has been locally removed by erosion, the bedrock surface is revealed. Blocks (Photo 2.4) that have emerged from the residue, form tors (Photo 5.42) when they accumulate in piles. Isolated blocks can sink down to the underlying bedrock and remain as balanced rocks (or rocking stones).

Mahabalipuram, near Madras, India

5.45 Valley in dry season
Valleys in the summer-rainfall areas of the warm-temperate subtropics and in the seasonally humid tropics are usually V-shaped valleys with floodplains. Water flow is periodic and low water or dry river beds reveal traces of fluvial activity (erosion and material transport).

Einasleigh River near Einasleigh, Queensland, Australia

5.46 Valley during flash flood
High water or floodwaters during flash floods are extremely effective geomorphological agents as they create a high-energy erosional environment. The river in the example was filled with rapidly-flowing water by heavy rainfall from a tropical cyclone. Even though the water level had already subsided considerably, two days after the precipitation event almost the entire valley floor was still flooded.

Burdekin River near Charters Towers, Queensland, Australia

5.47 Mountain Valley
In tropical rainfall zones, mountain valleys, such as those in the Andes, are deeply incised with steep walls and are therefore very narrow. They are V-shaped valleys and their floor is completely covered by the river bed. High perennial water flow rates and considerable transport energies result in high erosion rates. The view is from an altitude of 2400 m on to the approximately 400 m deeper valley floor of a sinuous incised valley. The photograph was taken from the ruins of an Inca settlement located in a saddle between two peaks of a granite complex.

Rio Urubamba on the eastern margin of the Andes near Machu Picchu, Peru

5.48 Block glide

Gravitational mass movements such as slides, slips, block glides, and sub-forest solifluction are common on slopes under humid tropical climate conditions. Lithological and structural factors on erosional scarp slopes can initiate large block glides with associated rock falls (Photos 2.17, 2.18). Limestone blocks of up to 15 000 m^3 have been involved in glide processes on the Lares erosion scarp in Puerto Rico. One of these block glides and rock falls moved a total material volume of 43 million m^3. The limestone blocks in the foreground are part of a rock fall that was activated by block glide processes and originated on a cavernous karst cone (Photo 7.15).

Lares erosion scarp, Puerto Rico

5.49 Termite mounds

As in the case of coral reefs (Fig. 29, Photo 10.25), termite mounds are organogenic microforms which are restricted to tropical areas. Termite mounds consist of highly resistant, cemented earth, plant, and faecal material and contain a maze of tunnels and chambers. They can reach a height of 7 metres and a basal diameter of 4 metres and are a frequent and specific microform in tropical savanna and forest areas.

near Hughenden, Queensland, Australia

6 Fluvial denudation landforms as a function of structure

Chapters 1, 2 and 5 contain repeated references to the fact that geological structure and lithological type – the two main endogenic factors – lead to distinctive weathering and erosion processes and thus to the development of different relief forms. This differentiation applies to all of the earth's climatic zones with their varying combinations of geomorphological processes. Structurally- and lithologically-dependent forms can be observed at all scales, i.e. as micro-, meso-, and macroforms. This chapter contains a systematic discussion of all landforms that show clear effects of geological structure and lithology type.

Micro- and mesoforms

Wherever rocks with different resistances are exposed to the effects of exogenic forces, differential erosion occurs: highly resistant rocks weather more slowly and are less likely to be eroded than less resistant types (Chapter 2, Photos 2.13, 6.1). The layering , i.e. the geological structure, of rock sequences with varying resistance is a decisive factor in relief formation. Resistance can vary vertically or laterally. The first case applies to all horizontally-bedded or only slightly dipping rock sequences. The second case applies to tectonically disturbed layers for which three basic options exist: 1. horizontal or only slightly dipping sedimentary sequences have been affected by block faulting, i.e. a block has been uplifted or downthrown relative to the adjacent block, resulting in resistance contrasts across the fault; 2. sediments have been folded leading to steeply dipping beds; 3. magmatic rocks have intruded into the sediments.

An example of differential erosion of horizontally-bedded sedimentary rocks is shown by the sculpturing of a structural bench in Photo 6.2. Alternating more or less resistant layers form stepped slopes down to lower surfaces (Photo 6.3). Steeply dipping beds, for example in fold or folded thrust sheet structures, form outcrops of particularly resistant layers that appear as ridges or hogbacks (Photos 6.4, 6.5). Similar forms resulting from differential erosion in less steeply dipping sequences can also be observed in cuesta (or homoclinal ridge or scarp ridge) relief. If magmatic bodies such as batholiths or laccoliths (Chapter 4) intrude into sedimentary complexes, they are frequently exposed by differential erosion owing to their greater hardness and resistance. The Mars Hill in northern New England (Photo 6.6) is an excellent example of a magmatic complex that has remained as a so-called monadnock on an eroded surface (Chapter 5). Magmatites in the form of sills and dykes that emanate from intrusive bodies along fractures to the surface, often appear as dyke walls or ridges. The most impressive example of this type of feature in Central Europe is the 150-km-long Bavarian and the 60-km-long Bohemian *Pfahl* (German: stake, post). This is a highly resistant, thick quartz vein which forms a vertical, 10–15-m-high dyke wall. The example in Photo 6.5 shows similar features to exposed magmatic dykes, but it is a sedimentary feature, formed by a thin sequence of vertically-inclined, highly resistant sedimentary layers.

Macroforms: Cuesta scarp, ridge scarp, and structural table land reliefs

Owing to their ubiquity, cuesta scarps (or escarpments) (Fig. 16) are one of the more important structurally controlled forms. Together with ridge scarps (see below), they define the morphology of the cover rocks on the unstable geological shelf (Chapter 3). Cuesta scarp relief can be subdivided into two basic, characteristic units: the cuesta scarp (or front slope) and the cuesta backslope. The scarp slope is the linear or slightly curved rock wall or front slope that has been created by erosion and denudation from a sequence of layers with varying resistance and mostly only slight dip angles. The scarp slope can usually be subdivided into a steep upper and a gentle lower slope, and at its top is a distinct edge called the crest. The piedmont angle marks the transition between the base of the scarp slope and the adjoining cuesta backslope. The upper scarp slope is formed in highly resistant rock, whereas the backslope often cuts across the highly resistant material to the less resistant rock which forms the base of the next step. In this case (Fig. 16a), the cuesta backslope cuts diagonally across the layers at a shallow angle. In arid zones the step surfaces mostly coincide with layer surfaces (Fig. 16b).

The forms of cuesta scarp relief can vary widely, for example in response to the relative thicknesses of the highly resistant rocks which form the relief step, and of the underlying less resistant rocks at the base of the relief step (Photos 6.7, 6.8). An additional striking feature is the top of the step slope (Photos 6.9, 6.10) which can be either sharp-edged or rounded (convex). In Fig. 16a, all of the step slopes are so-called front slopes, i.e. erosional scarps that are at opposite angles to the gen-

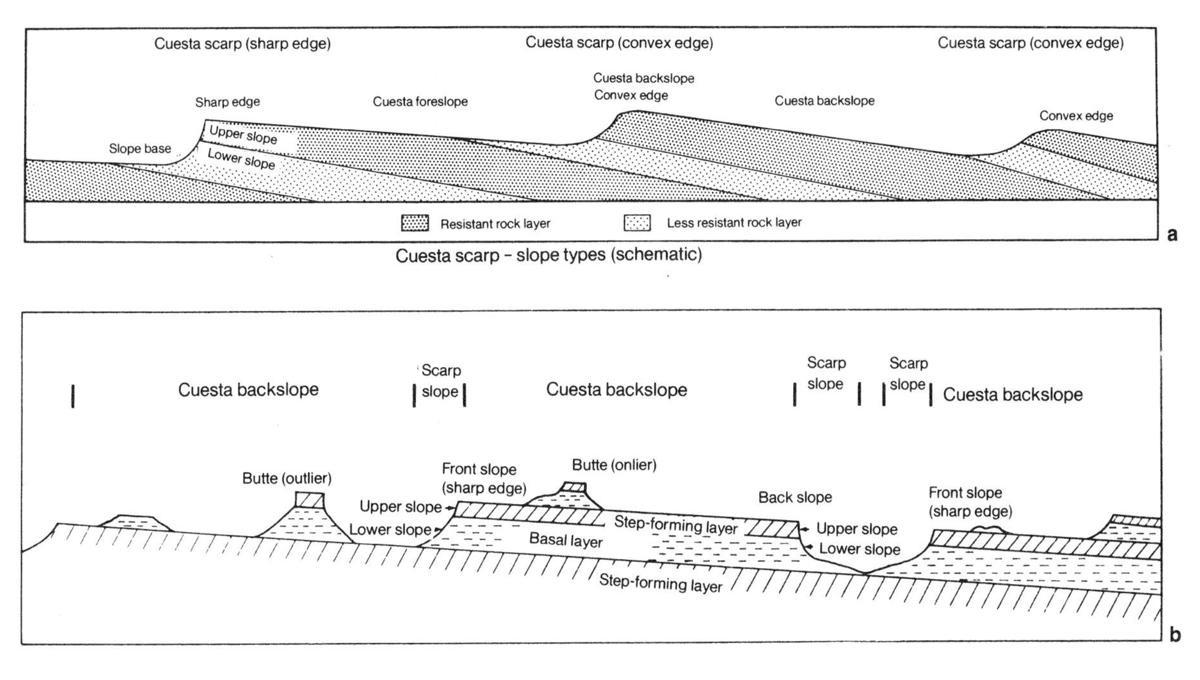

Figs. 16a, b Schematic profile of cuesta scarp relief
a Cuesta scarp - slope types (from Blume 1971) **b** Cuesta scarp relief in arid zones (from Schmidt 1988)

eral dip direction of the layers. Cuesta slopes that dip in the same direction as the tilted layers are called dip slopes or back slopes (Fig. 16b). The difference can be important, as erosional processes that are closely related to the spring line just above the base of the slope, are often stronger on back slopes. This is explained by the fact that both the slope face and the layers dip towards the back slope: both surface flow and groundwater flow therefore move in the same direction.

Present or past slope erosion and the resulting recession can be observed on both front and back slopes. The more resistant layer that forms the actual step is thereby gradually removed.

Slope recession rates can vary widely, even along the same erosional scarp: they can reach maximum values of up to 4 m per 1000 years, but are mostly less than 1 m per 1000 years. An example of erosional scarp recession resulting from block gliding on the slope base and rock falls from the upper slope is shown in Photo 6.11 from the frontal scarp of the Swabian Alb in Southern Germany. Slope remnants or spurs in front of the erosional scarp, which are still connected to the scarp base, and isolated remnants, so-called inselbergs (Photo 6.12), show clearly that slope recession is not always a linear, frontal process but can also appear as a broad erosional front in the scarp foreland. Removal of the step-forming rock sequence can also occur by surface wash or solifluction leading to gradual slope lowering or slope decline. Finally, a scarp can be removed from the rear by an advancing dip slope.

The diversity of geomorphological process combinations in different climatic zones can substantially modify the appearance of cuesta scarp relief. In the humid cool-temperate climatic zone, erosion and denudational slope degradation processes only play a minor role and the more aggressive erosional processes are rarely seen on scarp slopes (Photos 2.19, 6.11). Forms mostly reflect erosional processes that were active during the Pleistocene glacial periods. Convex slope tops (Photo 6.9) indicate solifluction activity, while sharp-edged slopes (Photo 6.10) can be explained by block glides over the basal layers of the scarp and by rock falls of the more competent scarp-forming layers. In arid zones, sheet wash processes during heavy rainfall are especially effective because of the sparse or non-existent vegetation. This is also due to the fact that the (cuesta) backslope surfaces mostly do not cut across layers, but instead coincide with layer boundaries. Intensive mechanical weathering and glide movements at the scarp base tend to form steep scarp slopes with sharp upper edges (Photos 6.13, 6.14). Frontal recession of the erosional scarp often coincides with the development of debris cones or ramps that are formed by the joint forces of sheet wash processes on the upper slopes and linear erosion in the lower zone of the scarp slope. They remain as triangular erosional remnants or spurs that mark the former locations of scarp slopes (Fig. 17, Photo 6.5). Both intensive recent erosion and erosional quiescence (Photo 6.16) can be observed on many scarp slopes in arid zones.

Steeply dipping layers, which commonly occur in anticlinal and synclinal structures (Chapter 3, Photo 3.5) on the unstable geological shelf, lead to the form-

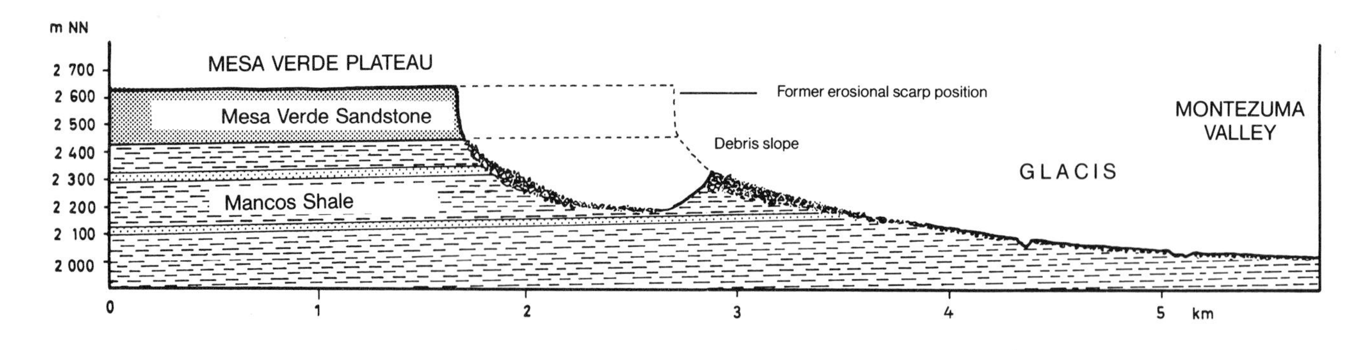

Fig. 17 Slope profile of the northern erosional scarp of the Mesa Verde in Colorado, USA, with debris slope (from Blume & Barth 1972)

ation of ridge scarps (or monoclinal ridge relief) instead of the typical cuesta scarp relief. As in the case of fold and thrust sheets (Photos 6.4, 6.5), selective erosion of resistant layers forms hogbacks, rock walls, and overhanging cliffs (Photo 6.17).

Horizontal layering of a sedimentary sequence with varying resistance leads to the formation of structural table land relief after erosional dissection. This type only occurs locally on the unstable geological shelf, but is quite common in the marginal areas of the continental shields that are covered by sediments, the so-called stable geological shelf (Chapter 3). The structurally-controlled, dissected peneplains with tabular erosional remnants (buttes and mesas) form the characteristic morphology of the structural table land relief. A classic example is provided by the canyon of the Colorado River (Photo 5.5) which has dissected the North American Colorado Plateau. In Central Europe the Elbsandstein Mountains (Photo 6.18) in Germany represent an imposing structural table land relief. The irregular distribution of erosional remnants of various sizes in Monument Valley on the Colorado Plateau (Photo 6.19) proves that the formation of structural table lands is not only the result of frontal scarp recession.

Landforms analogous to structural table land relief also occur in non-sedimentary rocks, for example, flat-lying magmatites, e.g. plateau basalts with intercalated volcanic ashes and tuff. The Western Ghats in India (Photo 6.20) are an excellent example, as their forms are virtually identical to those of structural table lands. Areas with lateritic crusts also form typical, tabular relief, as the relatively thin, surficial crust is mostly underlain by less resistant, weathered rock residue. In this case erosional processes are particularly aggressive because of the strong resistance variations. Rapid recent scarp recession and the resulting stepped slopes of the buttes form so-called breakaways (Photo 6.21).

6.1 Duricrust formation
Precipitated iron-oxides from iron-rich pore waters in sandstones can form hard crusts or duricrusts. The encrusted sections offer more resistance to weathering and erosional processes and remain as irregular rock formations.

Rock mountain near Duba (Dauba), Northern Bohemia, Czechoslovakia

6.2 Selective erosion (slope steepening)
Owing to their relatively high resistance to weathering and erosion, sandstone beds are often the cliff- and bench-forming layers in horizontally-bedded sequences of sedimentary rocks. The debris-covered slope sections with concave profiles mark the outcrops of less resistant layers. They can be serrated in climatic zones with strong sheet wash erosion.

near Longyearby, Spitsbergen

6.3 Selective erosion (slope steps)
Horizontally-bedded sedimentary sequences, composed of alternating resistant and less resistant beds, form step-shaped slope profiles. The example shows a concave basal slope area of less resistant marls. The overlying rocks are thickly bedded, Mid-Jurassic (Dogger) limestones with intercalated shaly layers. The resulting resistance differences are marked by alternating vertical and angled slope sections.

Western edge of the Causses near Millau, Massif Central, France

6.4 Hogback in thrust/fold structure
The Kalkvoralpen (Limestone Pre-Alps) in northern Austria to the west of the river Enns consist of rock sequences (thrust sheets) that have been overthrusted to the north. This type of deformation is a typical feature of the Northern Kalkalpen. Their relatively higher resistance leads the steeply-dipping Jurassic limestones in the example to form a sharp ridge and rock walls on the lower slopes. In the background (northeast) the rounded hills and valleys of the Flysch zone can be seen.

Gaishörndl (1100 m), south of Steyr, Austria

6.5 Rock wall

The rock wall in the example is formed by highly resistant Mid-Triassic (Muschelkalk) dolomitic limestones, which have been folded up into a vertical position and then eroded out of the surrounding less-resistant marls and shales. The resulting shape is similar to dyke walls formed by hard intrusive rocks (magmatites).

La Rochette, Bueche Valley northwest of Sisteron, Haute Provence, France

6.6 Monadnock

Monadnocks (or torso mountains) are hills or mountain chains which rise above a surrounding erosional surface because of their higher resistance to erosional processes. The monadnock in the example is a 550-m-high, isolated hill on the Aroostook Plain (150–180 m altitude). The surrounding area is a fossil peneplain (Photo 5.23) dissected by occasional valleys. The type locality of monadnocks is in New Hampshire, USA, where they are formed from remnants of a batholithic granite complex (Chapter 4).

Mars Hill, Northern Appalachian Mountains, Maine, USA

6.7 Erosional scarp (massive step)
The step is formed by the quartzitic, Cambro-Ordovician Bandiagara Sandstone that forms vertical or even overhanging cliffs as a result of its extreme resistance and great thickness. A sharp angle marks the transition to the underlying less resistant Koutiala Sandstone, which is covered by coarse debris.

Bandiagara scarp near Mopti, Mali

6.8 Erosional scarp (thin step)
The step-forming rock in this example is a platy, fractured sandstone band of the Carboniferous Millstone Grit which is exposed in a several-metres-high scarp and is underlain by thick shales. The entire scarp slope below the step is covered by coarse sandstone debris. The sandstone block in the foreground is capped by a water-filled weathered bowl (Chapter 2).

Pennines, west of Sheffield, England

6.9 Convex slope top

The front slope of this erosional scarp is formed by Cretaceous limestones (chalk) on the upper slope and shales on the lower slope area. Changing lithological types in the slope are clearly indicated by the different types of agricultural usage. Owing to the particular erosive conditions during the Pleistocene glacial periods (Chapter 5), a concave lower slope merges into a convex upper slope (waxing slope). Present climatic conditions in humid cool-temperate zones rarely lead to denudation processes. The parallel, wedge-shaped cuttings on slopes in the middle and background are old anthropogenic features and not natural erosional features.

Chiltern Hills near Luton, England

6.10 Sharp-edged slope top

In addition to convex slope tops, sharp-edged scarps also occur in humid cool-temperate climatic zones as the development of a specific shape is mostly controlled by lithological types. The example shows a butte formed by Carboniferous Dartry Limestones (caprock) underlain by Glencar Limestones. The steep slope is bounded at the top by a sharp edge which marks the transition to the table top. The debris-covered, concave lower slope of the butte consists of shales (Benbulben Shale).

Benbulben Range near Sligo, Ireland

6.11 Frontal recession of a scarp slope

The mechanisms that lead to frontal recession of a scarp slope are illustrated by the occasional block glides that occur on slopes of the Swabian Alb. The 30-m-thick Malm-β limestones that form the step are underlain by 180-m-thick impermeable shales. The shales tend to swell when moistened, and slip and glide processes are induced. This photo shows an overview of the block glide and associated rock fall in Photo 2.18. Following heavy rainfall, glide processes in the underlying shales as well as in the thick slope debris, reactivated old block glides that were formerly active during Pleistocene glacial periods and possibly also during the (subboreal) Holocene. 4 million cubic metres of rock debris and water-saturated shales were transported in a 400 × 1000 m area.

Hirschkopf near Mössingen, Swabian Alb, Germany

6.12 Inselberg

Hills or mountains that remain as isolated features in front of the actual erosional scarp and which consist of the same rock type are called inselbergs if all connections (spurs) from their base to the receding erosional scarp have been severed. They show that the step-forming rocks have been removed in the scarp foreland, i.e. between the scarp and the inselberg. This is frequently performed by frontal erosion of the scarp, but it can also be caused by other erosional processes (see Chapter 6). The step-forming layer in the example shown is a selectively-weathered Cretaceous carbonaceous sandstone, which frequently caps the marly base of the scarp slope.

Al-Qa'ara Basin in the Syrian-Iraqi desert-steppe plateau, north of Rutbah, Iraq

6.13 Front slope
Arid conditions generally lead to the formation of sharp scarp edges. The front slope of the erosional scarp in the example, with a step formed by the approximately 150-m-thick Jurassic Tuwaiq Limestones and underlain by less resistant marls in the slope base, has a sinuous course and forms a 300-m-high cliff above the foreland. Fresh scars on the vertical wall and channels on the steep, debris-covered slope base indicate strong, recent weathering and erosional processes. These are, however, only episodic occurrences.

Djebel Tuwayq, west of Riyadh, Saudi Arabia

6.14 Stepped slope
The erosional scarp shown here rises abruptly above the Wadi Tanezzuft to a height of up to 700 m. The base of the step is formed by the 100-m-thick, less resistant Tanezzuft Shales, which are overlain by the step-forming, resistant Silurian Acasus Sandstones. The scarp base is covered by a thick layer of freshly-deposited block debris. This feature, together with the deep incisions, illustrates the intensity of recent erosional processes. They are, however, only episodic, separated by long time intervals, under the present extremely arid conditions.

Adrar Acacus, Central Sahara, Libya

6.15 Erosional scarp (spur)

In all climatic zones, erosive dissection generally forms alternating scarp spurs and inlets along the back slopes. These forms dominate on the front slopes. The sharp-edged erosional scarp of the Mesa Verde is formed by the Cretaceous Mesa Verde Sandstone: the scarp base is the less resistant Mancos Shale. At a distance of about 1 km from the scarp front, a triangular talus flatiron marks a former scarp boundary location and illustrates the frontal recession process.

Mesa Verde near Cortez, Colorado, USA

6.16 Inactive erosional scarp

The erosional scarp of the Cretaceous Nubian Sandstone has many features which are indicative of intensive, earlier fluvial denudation, such as a sinuous course, well-defined slope benches, spurs, and inselbergs. These are completely missing at present: there are no fresh rock fall scars on the exposed slopes of the resistant layers, and no erosional channels on the stepped slopes. Block debris is not transported, as indicated by the presence of manganese coatings, and the slope base is concealed by the same finer-grained material that covers the foreland. The lithological boundary between the harder step-forming and the softer basal material, which is normally responsible for the morphodynamics of stepped slopes, has no effect in the example shown as the scarp base is covered by debris. Only aeolian erosional and depositional processes (Chapter 9) play a role, although it is a slight one in this example.

Djebel Ben Ghnema, Central Sahara, Libya

6.17 Ridge scarp

Ridge scarp relief (also called homoclinal ridge relief) is created by selective weathering and erosion of the outcrops of steeply-dipping layered sequences with varying resistances. The ridges follow the strike of the eroded layers. The Comb Ridge on the Colorado Plateau is formed by the Jurassic Navajo Sandstone. Highly resistant layers emerge as benches on the front slope, and can even be undercut, resulting in overhanging rock ledges. The resistant dolomites at the base of the front slope have been almost completely eroded and only chevron-shaped remnants are left (chevrons, Photo 2.25).

Front slope of Comb Ridge near Kayenta, Arizona, USA

6.18 Storeyed landscape

Varying erosional resistance in horizontally-bedded sedimentary sequences is often indicated by storeyed landscapes. The Elbsandstein Mountains consist of an alternating sequence of resistant Cretaceous sandstones and intercalated, less-resistant shales and marls. Isolated erosional remnants of the overlying sandstone beds are located on the erosional surface found at the top of the less resistant shales and marls. Valleys are frequently bounded by vertical cliffs, as they originally dissected the rectangular fracture system of the sandstones.

View from the Zirkelstein over the River Elbe towards the Schrammsteine, Sächsische Schweiz, Saxony, Germany

6.19 Erosional remnants (pillars)

In this example from the Colorado Plateau, the horizontally-bedded Permian De Chelly Sandstones that overlie the Organ Rock Shale tend to form steep cliffs but they have been virtually completely removed by erosion. Many erosional remnants are, however, spread over a wide area. Both the distance from erosional channels and the resistance differences in the sandstone account for differing erosional rates (Chapter 2). The erosional remnants therefore form large tabular blocks (mesas or buttes) or – in a more advanced stage of erosion – isolated towers and pillars.

Monument Valley, Colorado Plateau, USA

6.20 Storeyed landscape

The Western Ghats are composed of Trap, a 1–2000-m-thick sequence of lava (basalt) and volcanic ash. The high resistance of the basalts and the low resistance of the ash layers form storeys after erosional dissection of the original tablelands. Typical landforms of structural table land relief include structural surfaces, erosional remnants in the form of mesas, buttes and inselbergs, and benches and ledges on slopes.

Western Ghats near Kandala, India

6.21 Buttes (Duricrust cap)

Forms analogous to those developed in structural table lands can also develop in basement rocks if certain conditions are met. For example, if a lateritic crust has developed over a granite together with subsurface weathering, the crust can form erosional scarps. Below the frequently overhanging rock wall of the crust, weathered granite and blocks cover the slope base. Recent denudational processes controlling slope recession are extremely active. At the base of the slope, granite outcrops of unweathered bedrock emerge from the weathered and eroded debris.

near Meekatharra, Western Australia

7 Solution and precipitation landforms: karst

Processes by which rock-forming minerals pass into solution are one of the fundamental components of chemical weathering (Chapter 2). All rock types are to a certain degree soluble, but solution rates differ widely. For example in Central Europe, i.e. in a humid cool-temperate climatic zone with an annual mean precipitation rate of 800 mm, the following values were determined for mean surficial solution rates in mm per 1000 years (after Hohberger & Einsele 1979):

Gypsum, anhydrite	approx. 400
Limestone, dolomite	24–42
Sand, gravel, pebbles	0–36
Sandstone	2–14
Shale	6–8
Crystalline rocks	0–2

Gypsum and other salt minerals are so soluble that they do not occur on the surface under humid conditions. If they occur beneath the surface, their removal by solution forms caverns which can collapse and create hollows on the surface. While gypsum and salts are soluble in pure water, this process is much less effective for other rock-forming minerals. Their solubility is, however, much higher in acidic waters.

The most important rock solution process is the solution of limestone (main component: calcite, $CaCO_3$) by carbonic acid (H_2CO_3). This process is intensified by the purity of the limestone, i.e. by increasing $CaCO_3$ contents. Carbonate rocks are therefore not all as soluble as pure limestones. Carbonic acid converts the only slightly soluble calcium carbonate into the approximately ten times more soluble calcium bicarbonate:

$$CaCO_3 + H_2O + CO_2 + H_2CO_3 \rightleftharpoons Ca(HCO_3)_2$$

An additional factor is the effect of temperature on solubility: cold water can dissolve more carbonic acid than warmer water and the solubility of limestone is increased by higher carbonic acid contents. However, high carbon dioxide (CO_2) contents in ground waters are not the only important prerequisite for effective limestone solution: carbon dioxide contents of the troposphere (ocean waters above the thermocline) and of ground air are also critical factors. Carbon dioxide in ground air is generated by plants and bacteria and provides a supply to replace the CO_2 used during limestone solution. Frequently observed high CO_2 contents in soils on limestones are an important condition for intensive limestone solution. CO_2 contents in soils reach approximately identical values under different climatic conditions. Wide-ranging regional or local differences in the effectiveness of limestone solution (karstic corrosion) processes must therefore be explained by the amount of available water in the system – a function of climatic or soil characteristics. It is therefore understandable that the abundance of water in the humid tropics enables intensive limestone solution processes to take place.

Limestone is a very common rock type, and solution processes create specific landforms in all limestone areas. These forms are described as 'karst' from their type locality in the Slovenian mountain areas of Kras in the hinterland of Trieste where the first studies were undertaken. Both surficial and subsurface forms are generated. Surficial forms can be differentiated into covered and naked (open) karst, according to whether they are covered by weathered clay residue. In general, karst relief is typified by extreme morphological variety.

Solution forms (negative forms)

Microforms

Limestone solution (karst corrosion) generates many different forms of varying sizes. Impressive microrelief forms such as karren or rillenkarren (Photo 7.1) are often created on both naked and covered limestone surfaces. Larger exposed limestone surfaces can form a karrenfeld, while fractured horizontal surfaces can evolve into a so-called limestone pavement (Photo 7.2) with solution slots. Asymmetrical scallops (Photo 7.3) are common at high altitudes in alpine environments. Rounded shapes are often sculpted on soil-covered limestone surfaces due to slowly percolating waters containing humic acids. Sharp, fluted shapes which develop on exposed surfaces can later evolve into rounded shapes (Photo 7.4) when covered by soil. Fractures in limestones that are affected by solution processes also mostly show rounded surfaces on their walls (Photo 7.5).

Meso- and macroforms

Larger-scale solution cavities can be observed both on the surface of and within limestone bodies. They can be explained by specific hydrographic characteristics of the karst area. Limestone itself is a relatively dense and therefore mostly impermeable rock type, but fractures and bedding features commonly provide pathways for water flow. Surface waters percolate through fractures but also use specific karst forms such as pipes, dolines and sinkholes to penetrate to deeper levels. The waters are then channelled through extensive subsurface conduit systems which are often widened by corrosion to form caves. These cavity systems, which permeate the limestones, are hydrostatically pressured. However, a uniform water level does not exist and karst waters can move at different levels in adjacent conduit systems. In addition to local elevation differences, karst water levels in most areas are characterized by strong seasonal or short-term variations resulting from heavy rainfall. Two main hydrographic zones can therefore be differentiated in subsurface karst systems: in the upper, so-called vadose zone, karst cavities are only filled periodically or episodically, whereas in the deeper, phreatic zone, they are always filled.

If a limestone complex is completely or partially located at a higher elevation than valley floors in the surrounding areas, drainage by surficial water flow can be completely absent. Rivers that reach the karst can disappear into swallow holes (or ponors, Fig. 19). A classical example is provided by the Danube in the Swabian Alb near Immendingen in Germany. Underground water in karst systems can emerge from the limestones in marginal areas, and examples of karst springs in the Swabian Alb are the Aach Spring and the Blautopf (Photo 7.6), at a distance of 12 km from the respective swallow holes. Obviously karst springs can also only flow periodically or intermittently, namely when karst water levels are high.

In spite of the absence of surficial water flow systems, many karst areas are traversed by valleys. These dry valleys in karst areas (Photo 7.7) are relict forms, similar to dry valleys in arid zones (wadis; Photo 5.39). In Central Europe they were created under Pleistocene glacial climatic conditions. Karstic solution processes in permeable limestones generate many different types of cavities, such as karst pipes and dolines. Karst pipes or chimneys are vertical or subvertical cavities with widely-varying widths and depths which frequently mark intersecting faults. They can widen at depth and merge into caves or form karst 'wells' if they are closed at depth. Dolines are a

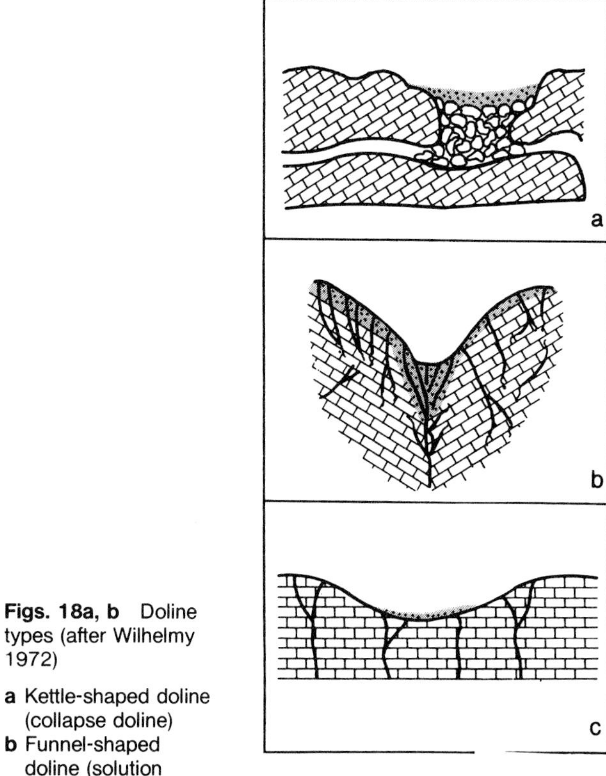

Figs. 18a, b Doline types (after Wilhelmy 1972)

a Kettle-shaped doline (collapse doline)
b Funnel-shaped doline (solution doline)
c Basin-shaped doline (solution doline)

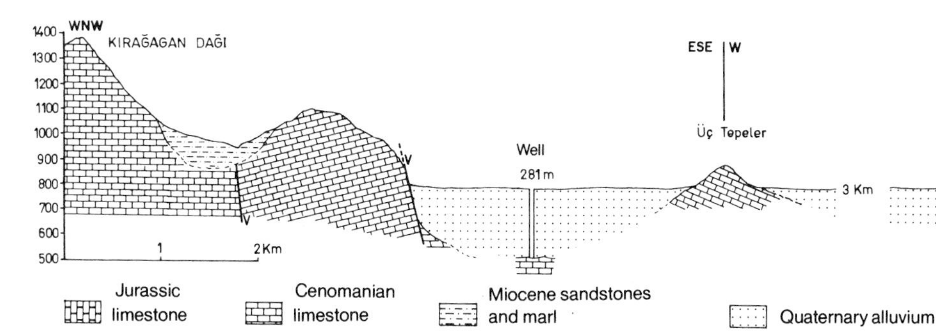

Fig. 19 Profile of the Kestel-Polje, Western Taurus, Turkey (from Guldali 1970)

Poljes (interior valleys) mostly occur in the synclinal structures of mountain chains and are particularly common in Mediterranean karst areas with this structural type (Photo 7.10). These karst forms have a flat valley floor (polje floor) of clayey weathered material that covers the underlying limestones and often reaches considerable thicknesses. A sharp boundary marks the edge of the polje floor where it meets the surrounding slopes. The formation of the level

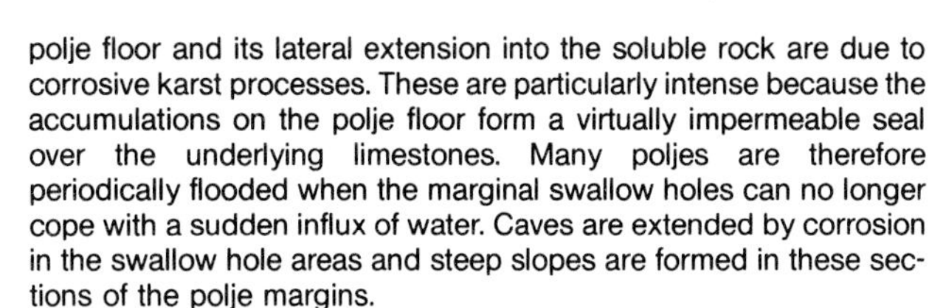
polje floor and its lateral extension into the soluble rock are due to corrosive karst processes. These are particularly intense because the accumulations on the polje floor form a virtually impermeable seal over the underlying limestones. Many poljes are therefore periodically flooded when the marginal swallow holes can no longer cope with a sudden influx of water. Caves are extended by corrosion in the swallow hole areas and steep slopes are formed in these sections of the polje margins.

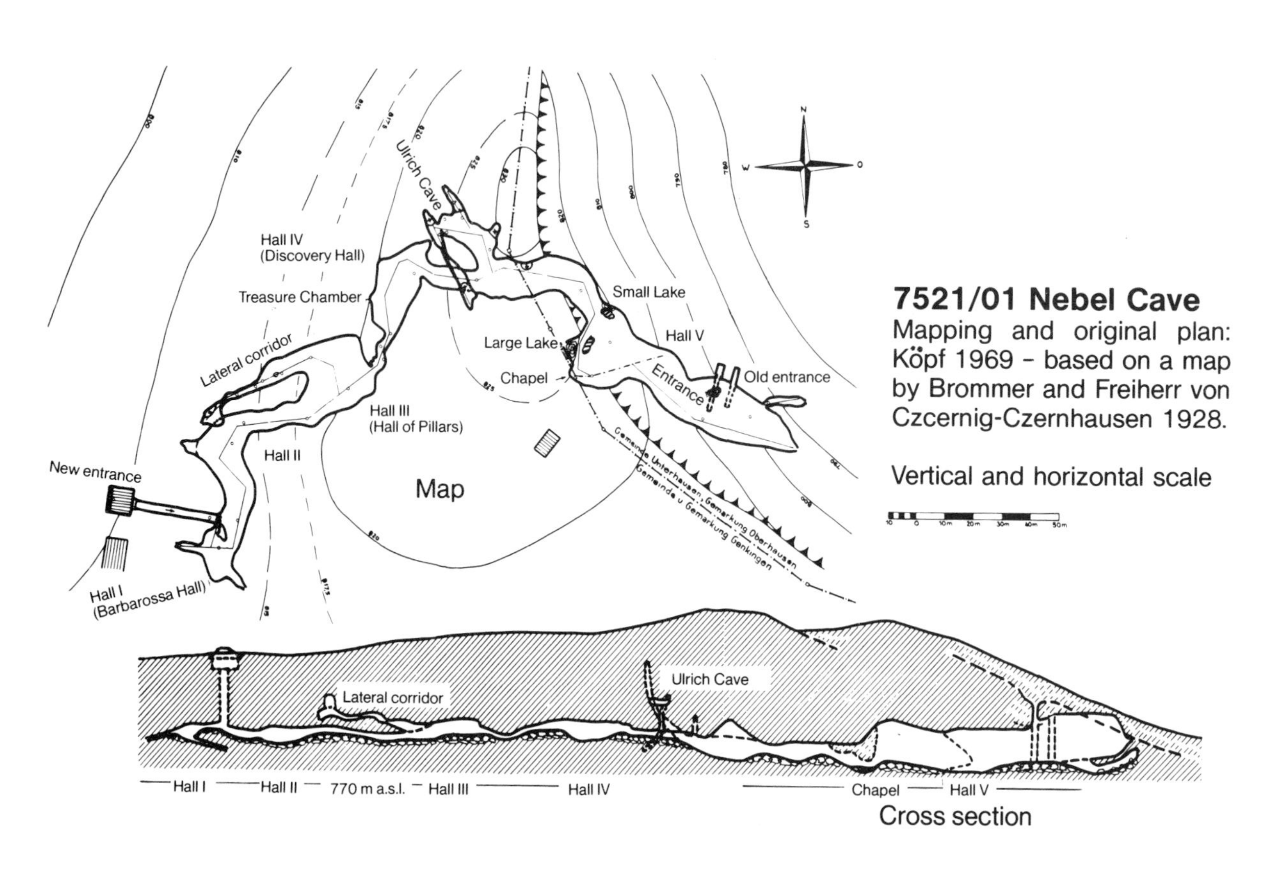

Fig. 20 Karst cave (after Binder, Bleich & Dobat 1984)

The Nebel Cave is located in one of the knolls of the Swabian Alb formed by sponge reefs of Middle and Upper Kimmeridgeian age. During the Earliest Pleistocene a river flowed through the cave which is now 65 m above the level of an adjacent dry valley. The river dried up when it was captured by other conduit systems due to continuing valley incision during the Pleistocene.

typical karstic form created by karst corrosion on limestone surfaces. Three basic forms can be recognized: kettle-, funnel-, and basin-shaped. Kettle-shaped dolines (Fig. 18a) are created by collapsing subterranean caves. Smaller forms in covered karst areas, which are usually funnel-shaped, are occasionally described as traps (Photo 7.9). Funnel-shaped dolines (Fig. 18b, Photo 7.8) are widespread and are regarded as the basic form of karst relief. Several small funnel-shaped dolines can be joined by corrasion processes to form basin-shaped dolines (uvalas; Fig. 18c, Photo 7.9). Funnel-shaped dolines show an especially wide range of sizes, and rows of smaller funnel-shaped dolines often mark fault lines. Poljes or interior valleys are large-scale depression features (Fig. 19, Photo 7.10), and can cover areas of more than 300 km². An example is the Livanjsko Polje (380 km² northwest of Split in Yugoslavia. Many poljes are flooded, mostly periodically, and permanent karst lakes can be formed if swallow holes are blocked. Clearing of swallow holes or the creation of new ones will subsequently drain the lakes. Karst lakes can also be formed when rising karst water levels fill dolines (Photo 7.11).

As previously discussed, limestone corrosion processes do not only create surficial karst forms - they are also responsible for subterranean forms. Even though surface waters that penetrate limestone formations lose much of their aggressiveness at the surface and in the vadose zone due to carbonate solution, sufficient carbonic acid contents still remain in karst waters to induce corrosion processes in the phreatic zone. This occurs when subterranean waters with varying carbonic acid contents and temperatures are mixed: carbon dioxide is released and the corrosive capabilities of the waters increase. The most impressive subterranean karst forms are karst caves (Fig. 20, Photo 7.18) with a widely-varying range of sizes and shapes. They represent extended conduit systems that can reach lengths of several kilometres, and can contain underground rivers or be partially and temporarily flooded. Mammoth Cave in Kentucky, USA, which is subdivided into five levels, is the largest explored cave system with a total length of 253 km.

Solution forms (positive forms)

Owing to the different solubilities of limestone and dolomite, selective karst corrosion occurs in partially dolomitized limestones. Karst relief is created which is characterized by positive relief forms (rock pillars), and which forms so-called block fields (Photo 7.12).

However, in addition to these locally restricted occurrences, many karst areas are characterized by positive solution forms instead of the previously described negative forms. These occur together with subterranean conduit and cave systems. Karst areas in which positive forms are the predominant surface feature are located in zones of particularly intensive limestone solution, with abundant water supplies and correspondingly high limestone solution rates. They are limited to the humid tropics (and occasionally subtropics). This positive karst form should not be described as 'tropical karst' in spite of its restricted occurrence in the tropics, as the ubiquitous doline karst forms also commonly occur in the tropics. The basic differences between both karst forms are occasionally characterized by defining doline karst forms as 'pock-marked', and positive forms as 'pimpled'.

Limestone pinnacles (Photo 7.13) created beneath a soil cover are a typical positive micro-form in humid tropical conditions. Characteristic mesoforms of positive exposed karst reliefs are knolls, cones, and pillars. These three karst relief types are usually classified together as cone-shaped karst. Knoll- or dome-shaped karst relief (Photo 7.14) is typical in thinner limestone formations, while cone-shaped forms (Photo 7.15) indicate deep-reaching corrosion processes. Laterally-spreading corrosion processes form haystack-shaped, steep karst hills (mogotes; Fig. 21, Photo 7.16) in knoll-shaped as well as cone-shaped karst reliefs. Isolated or grouped erosional remnants penetrate the surrounding areas, which are covered by limestone solution debris (marginal karst plain; Photo 7.17).

Limestone precipitation forms

Limestone solution is not the only creative chemical process in geomorphology; limestone ($CaCO_3$) precipitation from carbonate-saturated waters also creates micro-forms and occasional mesoforms such as speleothems (cave formations), sinter, and encrustations.

Speleothems in the form of stalactites and stalagmites (Photo 7.18) are characteristic features of karst caves. Loss of carbon dioxide from dripping water leads to limestone precipitation. Precipitation pinnacles (Photo 7.19) are a convergent form as they resemble karst pinnacles formed by limestone solution. Beach rock formation (Photo 10.8) is also caused by limestone precipitation. Evaporation of carbonate-saturated waters during capillary transport to the surface can generate surficial or subsurface limestone crusts (Photo 2.7). Carbonate-saturated water flowing over slope surfaces can also form limestone encrustations or even limestone sinter terraces (Photo 7.20) as a special type of sinter formation.

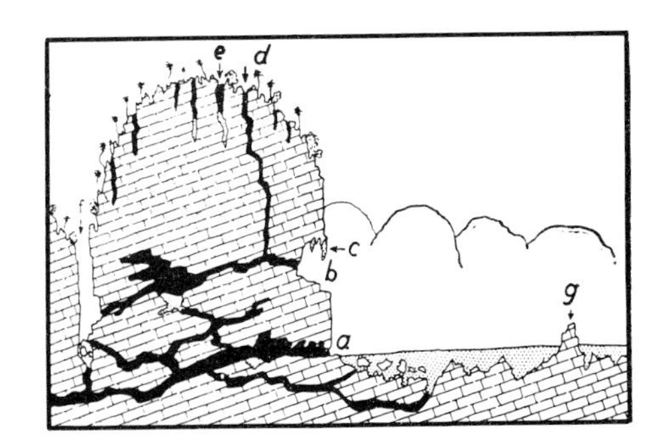

Fig. 21 Schematic section of a karst hill (mogote) in Cuba (from Lehmann 1953)

- a Basal cave with karren forms on ceiling
- b Semi-formed cave with stalactite curtain (c)
- d Karst pipe or chimney
- e Karst pipe, blocked by limestone solution debris
- f Karst 'passage'
- g Isolated limestone erosional remnant

Pseudokarst

Solution processes are not restricted to carbonate rocks. Other lithology types are also affected, although only to a much lesser degree, and the processes have a much smaller morphological impact than limestone solution. As usage of the term karst is generally restricted to forms that result from carbonate solution processes; solution forms in other rock types are described as pseudokarst. Solution of non-carbonate rocks is particularly evident on rock surfaces, on which karren (pseudo-karren) formation can occur which closely resembles that on limestones. However, subterranean cave systems can also be created. Pseudokarst forms in granitic rocks and sandstones are described as silicate rock karst (Photo 7.21), and their generation is also relatively strong under humid tropical/subtropical conditions, i.e. the same conditions which increase the intensity of karst corrosion processes, Pseudokarren can, however, also be found in other climatic zones and on other rock types (Photo 7.22). They can be particularly intense in coastal environments exposed to wave action (Photos 7.23, 7.24).

7.1 Rillen and fracture karren on exposed limestone surface
Rillen karren are solution forms with widths of several centimetres to decimetres (see hammer in photograph) that develop on sloping exposed limestone surfaces as a result of the solutional effects of flowing rain or melt waters. The ridges that separate the individual channels point downslope and are sharp, as rain water at these locations can dissolve more strongly than water already flowing down the slope. This is because increasing amounts of dissolved limestone reduce the solubility of the water. Fracture karren are also caused by solution processes. They form along fractures, can therefore be markedly straight, and can cross the slope of the rock surface. Rillen and fracture karren often occur together, as shown in the example of the Jurassic Quintener Limestone at an altitude of 2200 m. The fracture karren are mostly wider and deeper than the rillen karren because of a more intensive limestone solution. The cross-cutting fracture karren in the photograph are younger than the rillen karren.

Schrännen, Faulhorn Group south of Lake Brienz, Bernese Alps, Switzerland

7.2 Fault karren (limestone pavement)
Extensive limestone surfaces which are marked by karren formation form limestone pavements. Fault karren are the predominant feature on horizontal rock surfaces. The limestone pavement in the example is on a horizontal carboniferous limestone surface. The fault karren form a rectangular grid.

The Burren, County Clare, Ireland

7.3 Scallop-shaped karren
This micro-form is common on exposed limestone surfaces at high altitudes in alpine environments and can be explained by solution processes that occur under snow coverage. The scallops are larger on horizontal surfaces than on the sloping surface shown here. A prerequisite for the development of these forms is a moderate but constant water supply from melting snow. The scallop-shaped karren in the example are on the Jurassic Quintener Limestone at an altitude of about 2200 m.

Scrännen, Faulhorn Group south of Lake Brienz, Bernese Alps, Switzerland.

7.4 Rounded karren

Rounded or hollow karren are solution features that closely resemble rillen karren (Photo 7.1). The main difference is that the ridges which separate the hollows are not sharp-edged, as in rillen karren, but rounded. A common feature of the hollows is the basal lateral widening. Rounded karren are forms that have evolved from rillen karren owing to the solution effects of slowly-percolating waters with high humic acid contents. Soil cover is a precondition for this type of development and rounded karren are only found below the present tree line in alpine environments. The rounded karren in the example shown are from the Jurassic Quintener Limestone at an altitude of about 1400 m.

near the Brünig Pass, east of Lake Brienz, Bernese Alps, Switzerland

7.5 Fracture karren, formed below soil cover

Soil cover provides percolating waters with humic acid which produces rounded karren forms, as already shown in Photo 7.4. Limestone solution proceeds more rapidly because the humic acid and carbon dioxide concentrations in the soil are higher than on naked karst. Fracture karren of the size shown here with a depth of more than 10 m are described as karst 'passages'. They indicate extremely intensive limestone solution processes similar to those that occur under humid tropical conditions. The fracture karren in the example from the southern Provence on the surface of a Jurassic dolomitic limestone are relict forms.

Chaine de la Sainte-Baume, Provence, France

7.6 Karst spring

Water that disappears below the surface in karst areas re-emerges from limestones in karst springs. The drainage area of these springs is mostly quite extensive, leading to considerable flow rates. For example, the flow rate from the spring shown here (the Blautopf in southern Germany) averages 2000 litres per second. It usually varies between 50 and 4000 l/s, but can reach extreme values of 25000 l/s. The karst spring is a funnel-shaped bowl with a diameter of 40 metres and a depth of 21 metres, and is located at the upper end of an extensive, complex conduit system. The deep blue colour of the water is caused by its purity (1 mg/l suspended material), by the depth of the spring pond, and by the dissolved colloidal calcium carbonate. High flow rates lead to increased contents of suspended material, and the waters acquire a yellowish-brown colour.

Blautopf near Blaubeuren, southern Swabian Alb, Germany

7.7 Dry valley

Dry valleys in karst areas are relict forms. Two processes can have led to their formation: they were either formed when the present karst complex was not much higher than the surrounding valley floors, or they were created during permafrost phases which prevented surface waters from entering the limestones, so that surface water flow and fluvial erosion were the predominant processes. The latter explanation applies to all karst areas in cool-temperate climatic zones that were exposed to the effects of Pleistocene glacial periods.

Grosser Lauter Valley near Kohlstetten, Swabian Alb, Germany

7.8 Funnel-shaped doline

Funnel-shaped dolines (see also Fig. 18b) are the most common karst feature on both naked and covered limestone surfaces. The slopes of funnel-shaped dolines have moderate to steep angles and their size can vary considerably. A shallow debris floor is usually present, even in areas of naked karst. This is composed of limestone solution remnants that have been washed down from the slopes.

near Kosana, Kranjska (Krain), Yugoslavia

7.9 Basin-shaped doline with trap

Unlike funnel-shaped dolines, basin-shaped dolines (Fig. 18c) are usually only shallow depressions on soil-covered limestone surfaces. The typical oval, but also occasionally irregular, basin-shaped doline (uvala) can originate from the merging of a number of small funnel-shaped dolines that were located along fractures. The divides between adjacent funnel-shaped dolines and the originally steeper slopes are then flattened by corrosion. In the example shown, an oval basin-shaped doline in mid-Triassic limestones contains a small collapse doline formed by the collapse of the roof of a subterranean cavern (Fig. 18a). Collapse dolines in covered karst are described as traps.

Sindolsheimer Höhe near Altheim, Badisches Bauland, Germany

7.10 Polje

Poljes are completely enclosed depressions (interior valleys) that occur as meso- or macroforms in karst areas (Fig. 19). The polje shown here is in a thrust sheet area of southern Provence in France. The depression has a length of 5 km and a width of 2 km and is formed in soluble Jurassic and Cretaceous limestones. Water that accumulates above the more than 40 m thick clayey weathered material during the winter rainfall season cannot be drained rapidly enough by the swallow holes along the margin of the polje, and the polje floor is therefore periodically flooded.

Cuges, south of the Chaine de la Sainte-Baume, Provence, France

7.11 Karst lake
Karst lakes form periodically in poljes (Fig. 19, Photo 7.10) and only occasionally survive for longer periods. However, they can also occur in dolines filled by rising karst water levels. This is the case in Florida, where dolines were created in Miocene limestones during the Pleistocene glacial periods when sea water levels were much lower. Post-glacial eustatic sea level rise then raised the karst water level and created lakes in the dolines. The so-called Lake District in Florida contains many karst lakes of varying sizes.

near Orlando, Florida, USA

7.12 Block fields
As in the case of massive, fractured sandstones (see Photo 2.12), rock pillars can also occur in limestones and form block fields. Partial dolomitization of limestones is responsible for the creation of the block field shown here. As the dolomitic parts of the limestone are less soluble than the pure limestone areas, they are less strongly affected by karst corrosion. This example was created during the Pleistocene.

Montpellier-le-Vieux, Causse Noir, Massif Central, France

7.13 Limestone pinnacles
In areas of particularly intensive karst corrosion, under humid-tropical climatic conditions, limestone pinnacles develop as a characteristic micro-form in karst areas. They are created below a soil cover. Limestone is removed by continuous widening of criss-crossing karst passages (Photo 7.5) until only individual pinnacles remain as erosional remnants.

Dong Lan area, northwestern Khorat Plateau, Thailand

7.14 Karst knolls

Knoll- or dome-shaped karst relief that develops from thinner limestone formations, can be observed in the tropics. It is one of the typical tropical cone-shaped karst forms. Large groups of knolls with evenly rounded dome shapes and similar heights were first observed in Java and were defined as the Gunung Sewu type of cone-shaped karst. The 70–100-metre-high karst domes in the example are morphologically identical to this type. However, several features are not typical for Gunung Sewu karst, such as the intercalated basins and marginal plains, and the occurrence of karst springs at the base of the knolls. The knoll-shaped karst relief on Bohol is restricted to synclinal structures with horizontally-bedded Pliocene and Pleistocene limestones. The ranges that surround the knoll-shaped karst areas consist of the same limestone formations and form anticlinal structures with dip angles of up to 40°. Completely different karst forms have developed in these areas. The perfect symmetry of the karst domes on Bohol is the result of uniform lithological structure. The uppermost units are more resistant than the basal units. This also explains the overall similar elevation of the domes.

Bohol Island, southwest of Carmen, Philippines (photograph by U. Scholz, Giessen, Germany)

7.15 Karst cones

Cone-shaped karst relief is created by deep-reaching corrosion, resulting in increased potential energy compared to the knoll-shaped karst areas. Between the erosional remnants of the karst cones, corrosion processes form hollows that are deeper than those associated with doline formation. Lateral solution processes, which widen the base of the hollows, are active at the impermeable base of the funnel-shaped corrosion forms. This leads to steep-walled cones as well as to the commonly star-shaped outlines – unlike the mostly rounded dolines. These features are described as cockpit karst on Jamaica. The karst cones in the example are on the southern margin of the cone-shaped karst area, and are composed of the (up to) 300-m-thick Oligocene Lares limestone.

near Caguana, Puerto Rico

7.16 Karst hill

Haystack-shaped karst hills (mogotes) are erosional remnants of extended cone-shaped limestone complexes and indicate a late stage of karst corrosion. They rise abruptly as isolated features or in groups from surrounding small plains that are covered by limestone solution remnants. They have developed by lateral karst corrosion from cockpits and marginal karst plains (Photo 7.17) into the karst areas. These karst hills, which are described as mogotes (Fig. 21) from their type location in the Sierra de los Organos in Cuba, are covered by mostly sharp-edged karren and are permeated by karst chimneys and caves. External shallow caves (balms) on the hillsides mark previous karst water levels which controlled lateral karst corrosion. Stalactite curtains across the entrances of external caves and karren marks on the cave ceilings are indicative of strong solution/precipitation processes under humid tropical-subtropical conditions. The mogote hill shown here, together with others visible in the background, form a line of hills along a tectonic fault zone. The widespread haystack-shaped karst relief in the Guilin area is predominantly formed in Devonian limestones. The karst hills rise above the surrounding plains to a height of more than 200 m.

near Guilin, Guangxi Province, People's Republic of China

7.17 Marginal karst plain

Plains that extend from the edges, but also exist within, karst hill areas are formed by corrosion, similar to the polje valleys in doline karst areas (Photo 7.10). Strong lateral karst corrosion is directed from the plains that are periodically flooded under the typically humid-tropical climatic conditions. Basal caves (Fig. 21) are commonly formed in the karst hills and their collapse serves to maintain the steep flanks of the haystack-shaped limestone erosional remnants.

Valle de Vinales, Sierra de los Organos, Cuba

7.18 Stalactites and stalagmites (speleothems)
The constant dripping of carbonate-saturated water in karst caves and the concomitant loss of carbon dioxide leads to the precipitation of calcium carbonate in the form of stalactites suspended from the cave ceilings, and stalagmites that grow upwards from the cave floors. Stalactites and stalagmites are made of concentric layers of calcium carbonate and can join to form limestone columns. Considerable variations in the size and shape of speleothems are possible.

Bear Cave near Sonnenbühl-Erpfingen, Swabian Alb, Germany (photograph courtesy of Landesbildstelle Württemberg, Stuttgart, Germany)

7.19 Precipitation pinnacles
Precipitation pinnacles are a convergent form to karst pinnacles formed by limestone solution (Photo 7.13). In the example shown here, pinnacle sizes range from several centimetres to 2 metres. They were created by a polygenetic process. In phases of eustatic sea-level lowering during the last Pleistocene glacial period, beach sands with high contents of shell fragments were blown inland to form dunes. During later more humid phases, the dunes were covered by vegetation, and carbon dioxide-rich, downward-percolating waters dissolved the calcium carbonate in the shell fragments. Limestone precipitation in the form of vertical concretions of cemented sand grains occurred above an impermeable base in the dunes. At present, sand is being removed by wind transport under arid climatic conditions, revealing the limestone concretions which are then exposed to weathering and erosion.

Nambung National Park, between Geraldton and Perth, Western Australia

7.20 Sinter terraces
Flow on sloped surfaces can increase evaporation and carbon dioxide emission from calcium carbonate saturated waters, and thereby induce limestone precipitation. A stepped series of sinter bowls can result from the precipitation – so-called sinter terraces. Turbulent flow over the edges of the bowls leads to additional loss of carbon dioxide and calcium carbonate precipitation, and horizontal growth of the sinter terraces. Limestone precipitation and terrace growth is particularly rapid if, as in the case shown here, water is supplied from hot springs. Sinter encrustations have not only covered the buildings of the ancient city of Hierapolis, but have already covered recently constructed irrigation channels used for agricultural purposes in the region. The sinter terraces cover a steep, approximately 50-m-high slope, and are several hundred metres wide.

Pamukkale, near Denizli, Turkey

7.21 Karren in silicate rocks
Sandstone surfaces in humid cool-temperate climatic zones can be marked by karren in addition to typical features such as solution bowls (Photo 6.8). Rillen karren can be observed on rock surfaces as well as bumpy, cup-shaped micro-forms. In the example shown, they have developed on fractured Cretaceous sandstones. Large-scale fracturing indirectly controls the development of funnel-shaped collapse structures that form at fracture intersections and resemble those formed as collapse dolines in limestones.

Schrammsteine, Sächsische Schweiz, Saxony, Germany

7.22 Fault karren on laterite crust
The surface of lateritic crusts can be marked by solution forms in the shape of fault karren. They can be found on the upper edges of buttes, where contrasting rock resistances between the crust and the underlying weathered material, together with fracturing of the crust, cause active slope recession.

near Greenvale, Queensland, Australia

7.23 Rillen karren on silicate rocks
In coastal environments with direct exposure to wave action, and especially in areas with particularly high tides, solution processes in run-off waters lead to the development of rillen karren. In the example shown, chemical solution processes are the controlling factor for typical micro-forms developed in granites on the French Atlantic coast.

near Tregastel, Brittany, France

7.24 Cavity formation
In addition to karren formation, cavities are a common form in wave-exposed areas. In the example, cellular erosion cavities mark a cliff composed of Ordovician shales. Cavity formation with a wide range of sizes and shapes (see lens cap for scale) results from chemical solution processes similar to those involved in karren formation.

Pointe de Trefeuntec, Bay of Douarnenez, Brittany, France

8 Glacial landforms

Glaciers

Glaciers cover about 15 million km², i.e. about 10 per cent of the mainland surface area of the earth (ice-covered areas of the oceans not included). Their extent was considerably larger during the Pleistocene glacial periods. As already described for fluvial processes (Chapter 5), glacial processes can create both erosional as well as depositional forms. However, many of these are only exposed when the glacial ice cover melts.

Glaciers are ice masses created from snow by the compressive forces of overlying thick snow masses. This compression leads to considerable density increases. Freshly fallen snow has a density of 0 · 1 to 0 · 3, firm snow 0 · 55 to 0 · 75, firm ice 0 · 8, and glacier ice 0 · 9 g/cm³.

Ice masses are described as glaciers when they start to move according to the force of gravity. Glaciers can be subdivided into accumulation and destruction (ablation) zones. In the first case the accumulation rate exceeds the ablation rate, while in the latter case the total mass of ice is being reduced. The theoretical boundary line between accumulation and destruction areas, i.e. where accumulation equals loss, is defined as the equilibrium line.

Various criteria can be used to classify glaciers, such as glacier ice temperatures, the type of glacier movement, or relief-related shapes. Owing to the constant low temperatures, polar (cold) glaciers at high latitudes have no meltwater drainage systems. Meltwater drainage is, however, a characteristic feature of all glaciers in other regions – the temperate (warm) glaciers. Glaciers move as a function of slope angle either in gradual flow movements or as block movements. In the following text, the criterion of glacier shape is used in the discussion of glacier types. Other important characteristic features that are used to classify glaciers are, however, also mentioned. Glacial areas exist in which glacier forms are subordinate to the local relief, as in the case of valley glaciers. This type is mostly exemplified by relatively small glacial areas at medium and low latitudes. On the other hand, the inland ice areas of Greenland and Antarctica are covered by extensive ice masses that completely cover the pre-glacially-formed relief. In these cases glacial processes control relief formation.

The best known type of relief-dependent glacier in Europe is the valley glacier (Photo 8.1). It represents the specific glacier form at high altitudes in the non-polar areas and is therefore often described as the alpine type. Most valley glaciers accumulate in cirques (cwm, corrie; Fig. 22, Photos 8.2, 8.11). The ablation zone at the lower end of the glacier, the glacier tongue, extends into valley areas. Temperate valley glaciers have widely varying lengths, with maximum values of 25 km in the Alps, while polar glaciers can exceed 200 km in length. Valley glaciers can merge to form piedmont glaciers (Photo 8.4) when they emerge from highland areas. Additional forms of relief-dependent glaciation are cirque glaciers (Photo 8.2), and ice caps formed on mountain peaks in the tropics

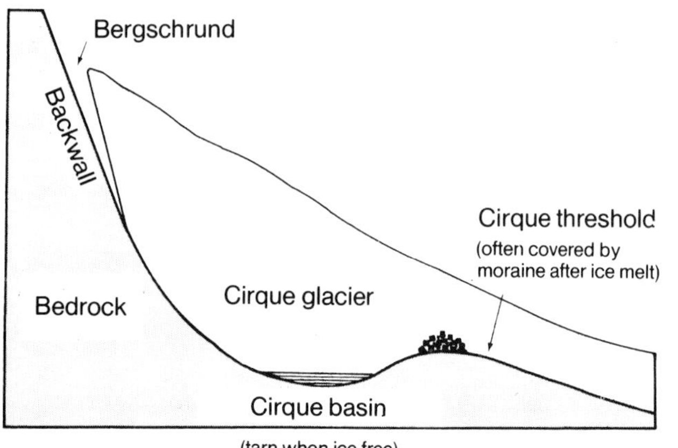

Fig. 22 Cirque, schematic section

Cirques (Photos 8.2, 8.11) are hollow forms that have evolved from slope depressions or spring hollows as a result of glacial erosion. Snow accumulates and is converted to ice in the hollows. Cirques are therefore the original locations of glacier formation and frequently represent the glacial source area. Typical morphological features are irregular but mostly semi-circular plan form, steep backwalls, bowl-shaped excavation of the cirque basin, and frequent frontal closure by a cirque threshold or ridge. A typical feature of ice-filled cirques is the deep crevasse (*bergschrund*) at the rear of the cirque, between backwall and ice, caused by different heat absorption rates between rock and ice in summer. Repeated melting of the light-coloured snow (*firn*) next to the dark rock wall accelerates ice formation. The ice itself – with the effect amplified by frost weathering – steepens the backwall by plucking processes. Rock fragments loosened by frost weathering fall from the rock face on to the glacier surface. The ice mass together with the included rock fragments moves along the cirque base and excavates the characteristic bowl-shaped cirque by abrasion and plucking.

(Photo 8.3). Transitional forms between glaciers and inland ice masses, which control relief formation, are (trunk) glacier networks (Photo 8.4), which leave areas of exposed rock.

Glacier movements form crevasses or fissures in the glacier ice mass. Ice flow rates in the centre of the glacier higher than those on the flanks lead to the formation of longitudinal crevasses. Cross-cutting crevasses are caused by changing slope angles of the glacier base. Glacier surfaces which are cut by intersecting deep crevasses, or are very steep (icefalls), or which are broken into a chaotic mass of broken blocks, are called seracs (Photo 8.14). Radial crevasse systems (Photos 8.5, 8.13) are characteristic features of glacier tongues with reduced thicknesses and increasing widths. Calving occurs when floating glacier tongues break up to form icebergs (Photo 8.6).

Many different micro-forms are created on the glacier surface during ablation. One such form is hollows (cryoconite holes; Photo 8.8) caused by rock particles that absorb solar radiation and sink into the surrounding ice. Larger boulders can form temporary glacier tables on the ice surface by protecting it from ablation. Stagnant debris covered glaciers can develop forms described as glacial 'karst' or 'thermokarst' (Photo 8.7) as a result of selective ablation processes. These processes also lead to the preservation of ice remnants - dead ice (Photo 8.8) - that are left by retreating glacier tongues.

Glacial erosional processes and forms

Glacial erosion processes can be subdivided into three main types: grinding or polishing of rock surfaces (abrasion), removal of material from rock surfaces (plucking), and removal and transport of material in basins formed by the glacier tongue. The effectiveness of ice erosion can be considerably increased by fluvioglacial forces from the subglacial meltwaters that exist below subpolar and especially temperate glaciers. Moving ice can polish rock surfaces which are then scratched or striated by rock fragments embedded in the ice (glacial striation; Photo 8.9). Abrasive action by moving glacier ice can form rounded exposed rock surfaces, so-called rock drumlins or *roches moutonnées* (Photo 8.10). The formation of cirques and U-shaped valleys can be explained by both abrasion and plucking processes. Initial glacier formation occurs in cirques (Fig. 22, Photos 8.2, 8.11), which are bowl-shaped hollows with steep head walls that are located on slopes. Cirques are excavated and deepened by glacial erosion. U-shaped valleys (Photo 8.12) are valleys that have been remodelled by valley glaciers (Photo 8.1). As in cirques, valley slopes are steepened by glacial erosion. Considerably greater ice thicknesses in the trunk valleys excavate them more deeply than tributary valleys. After the ice has retreated, tributary valleys remain as 'hanging valleys' at higher levels than the main valley floor. This abrupt step (confluence step) between trunk and tributary valley results in the tributary river entering the trunk valley by way of a water fall or in a deeply incised gorge.

On the other hand, if a glacier enters a tributary valley from the main valley, the total mass of ice, and therefore the erosive force of the main glacier, is reduced. A diffluence step is created in the main valley, which rises downstream, and can lead to the formation of a lake in the valley after the ice cover has melted. Fluvial incision by postglacial water flow can cut through the step, forming a narrow valley. Glacial tongue basins are also deepened by glacial erosion. Plucking processes on underlying bedrock play a role as well as erosion of detrital rocks by subglacial meltwaters. After the ice cover has melted, the glacial tongue basins surrounded by terminal moraines are mostly filled by proglacial lakes, as exemplified by Alpine foreland areas and the moraine relief of Northern Germany. Flooding of the sea into these glacier valleys can form a fjord (Photo 10.14).

Glacial depositional processes and forms

Rock debris that is transported and deposited by glaciers is defined as moraine material, and is characterized by an extremely heterogeneous mixture of widely varying grain sizes, sorting, and roundness. Larger rock fragments are rounded and striated as a function of transport distances. The typical light grey colouring of glacier meltwaters (glacier milk) caused by suspended material is a clear indication of the large per-

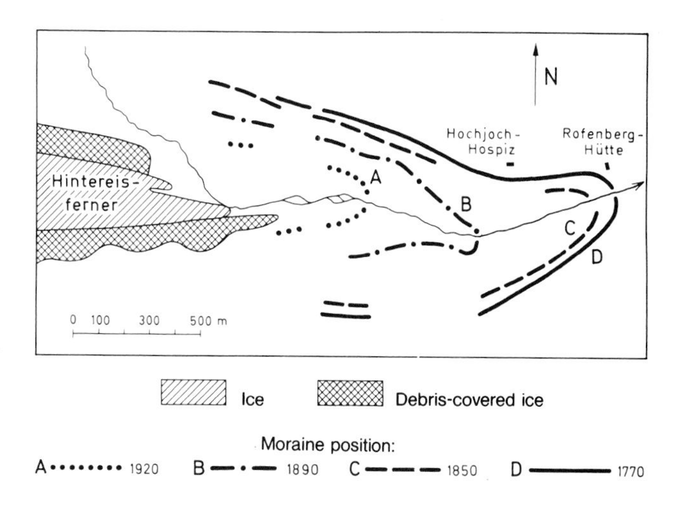

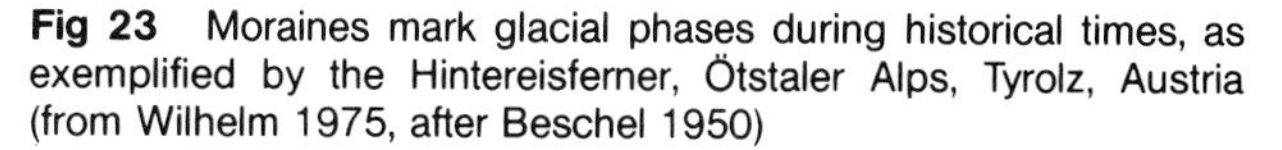

Fig 23 Moraines mark glacial phases during historical times, as exemplified by the Hintereisferner, Ötstaler Alps, Tyrolz, Austria (from Wilhelm 1975, after Beschel 1950)

Four glacial advances between 1770 and 1920 are clearly indicated by the walls of terminal and lateral moraines. The glacier front has retreated by 1 km over this period. Moraines mark interruptions in the overall glacial retreat by brief advance phases. Glacial retreat is still continuing at present.

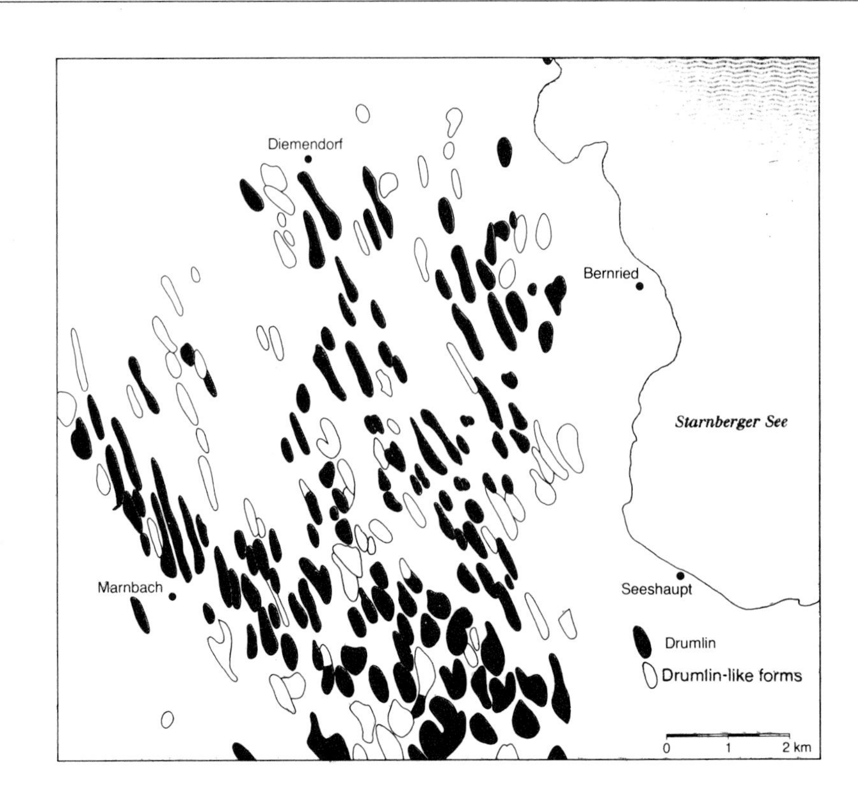

Fig. 24 Eberfinger Drumlin Field (from Petermuller-Strobl & Heuberger 1985)

A drumlin is an elongated 'streamlined' glacial deposit with a generally asymmetrical, elliptical basal shape. The highest point of a drumlin is usually close to the blunt end, which faces into the glacial movement direction, while the end that points in the direction of glacial movement has a more gradual slope angle. Maximum dimensions are lengths of several kilometres, widths of 1000 metres, and heights of 70 metres. They are usually located close to the terminal moraine in the ground moraine zone, and their long axis is parallel to the direction of ice movement. Drumlins rarely occur in isolation and typically form large groups or drumlin fields. This is clearly visible in the Eberfinger Drumlin Field, that was formed during the last ice age in the ground moraine area of the Isar-Loisach glacier between the glacial tongue basins of Lake Starnberg and Lake Ammer in Southern Germany. The subglacial formation of the drumlins from ground moraine material can be explained by lower ice movement velocities in the inner marginal glacier zone. The drumlins in this figure are geomorphologically defined as ridges with smooth streamlined shapes and heights of at least 7 metres. Drumlin-like forms are those which partially fulfil the previously-mentioned critieria. They represent transitional forms to uneven ground moraine landscapes.

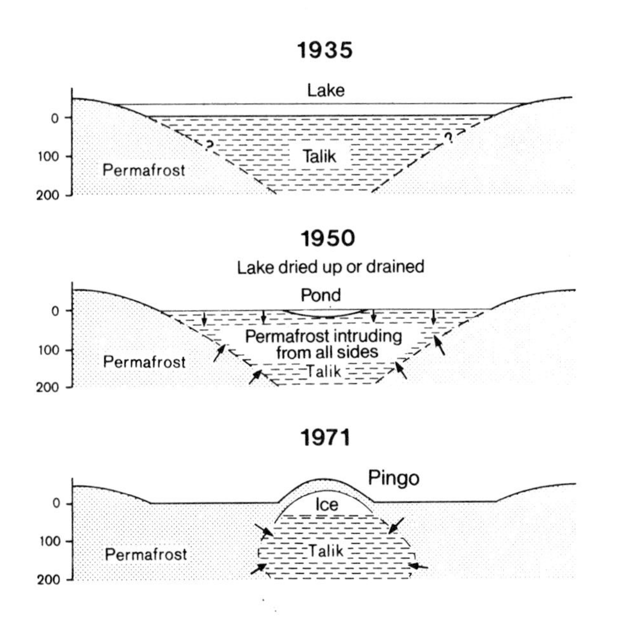

Fig. 25 Pingo (from Schultz 1988, after Mackay 1972)

Pingos (Photo 8.20) are forms that occur in non-glaciated, i.e. periglacial areas of subpolar climatic zones. Their development is related to permafrost and frost-dynamic processes. Pingos are cone-shaped hills with an ice core, a height of up to 100 metres, and basal diameters of 300 to 1200 metres. They occur on subpolar tundra plains, especially when permafrost becomes thin or patchy. A prerequisite for their formation is the presence of a lake on a non-frozen subsurface zone (talik). If sediments fill the lake, the entire water depth can freeze in winter, leading to ice formation in the upper levels of the wet talik. The permafrost zone can then expand into this area, leading to increased hydrostatic pressures in the talik, which push up the overlying sediment and ice layers. Growth rates of pingos can attain 1 · 50 metres per year during the initial growth phases, but then decrease rapidly. Particularly large pingos can reach an age of about 1000 years, but continuing pingo growth exposes the ice core to ablation processes, and the pingo then disappears. The frost-dynamic processes of pingo formation can then re-commence.

centage of very fine grained material in moraines. Various moraine types can be differentiated, such as ablation, lateral, medial, terminal, and recessional moraines. Ablation moraines (Photo 8.15) consist of rock debris that is carried on the upper surfaces of glaciers; lateral moraines (Photo 8.14) form on the sides of glaciers; medial moraines (Photo 8.1) are created when glaciers unite; and terminal moraines (Photos 8.13, 8.16) are deposited at the front of glaciers. Lateral and terminal moraine walls often mark previous glacier limits, and enable the determination of phases of glacial advance and retreat (Fig. 23). Older lateral moraines are generally not clearly recognizable in the field. Material transported at the base of glaciers forms ground moraines (Photo 8.17). These can be deposited as even thin layers or as knoll-shaped ground moraines. An even ground moraine is a characteristic feature in areas in which ice has been removed by the melting of static glaciers on relatively level ground. Knoll-shaped or hilly ground moraines are found in areas with oscillating glacial retreat phases. A typical form in the terminal moraine zone of ground moraine landscapes are drumlins (Fig. 24) that resemble large groups of streamlined humps. Occasional isolated erratic blocks (Photo 8.18) are located in ground moraine zones. These are large rock fragments that can reach considerable sizes and have been deposited at some distance from their original outcrop. Hollows (dead ice kettles, Photo 8.19) can be created in ground moraines from dead ice occurrences that were left behind by melting glacial ice. Additional ice-induced structures are so-called pingos (Fig. 25, Photo 8.20) that occur in subpolar periglacial areas as ice-filled debris hills.

Fluvioglacial and glaciofluvial forms

Both of the above expressions are used for forms created by meltwaters; those created in glaciated areas are called fluvioglacial, while those that develop in the glacial foreland are described as glaciofluvial. Erosional and depositional forms must be differentiated. Some typical erosive fluvioglacial micro-forms are glacier mills on the ice surface, and potholes (Photo 8.21) that are abraded into the solid bedrock by the grinding effects of rock fragments in turbulent currents below crevasses. Fluvioglacial meso-forms can be observed in the terminal moraine zone in glacier tongue basins: tunnel valleys are created by erosion in subglacial meltwater streams. These appear as long narrow lakes after the ice cover has melted. Fluvioglacial accumulations form wall-shaped ridges, so-called eskers (os; Photo 8.22). They mark the courses of subglacial meltwater channels, in which deposition occurred under hydrostatic pressure. Glaciofluvial processes along the outer margins of glacial areas are mainly controlled by depositional activities. Meltwaters deposit alluvial fans (Chapter 5) in the form of outwash fans (sandur; Photo 8.23) in front of the terminal moraines. They can cover large areas, especially along the margins of extensive inland ice sheets, such as those that existed in Northern Germany.

8.1 Valley glacier

Valley glaciers move in valleys formerly excavated by linear fluvial erosion. Valley glaciers mostly flow slowly and smoothly, a type of movement that is typical for viscous masses. The middle of the glacier moves more rapidly than the sides. Gradient steps can, however, change the type of glacier movement, and block movement then occurs (seracs; Photo 8.14). The photograph shows how tributary glaciers merge into trunk glaciers. The lateral moraine of a tributary glacier continues in the trunk glacier in the form of a medial moraine.

Mount Denali (6193 m, formerly Mount McKinley), Alaska, USA

8.2 Cirque glacier

Cirque glaciers are small glaciers that fill hollows (cirque; Fig. 22, Photo 8.11) on mountain slopes. The cirque glacier shown here is located at 79°35′ North. It is a polar glacier and the process of compaction from snow to firn to ice takes several years. On the left hand side the steep-walled calving cliff of the Waggonway Glacier can be seen, which received its name from its abundant crevasses.

Southern side of Magdalena Bay, northwestern Spitsbergen

8.3 Ice cap

Smaller ice fields, especially those that occur on high mountain peaks in the tropics, are called ice caps. They are typically well formed on volcanic structures. The elongated Cordillera Blanca in Peru, formed by a granitic intrusive body, has many ice-covered peaks. The highest is the Nevado Huascaran (6768 m) shown here. Glaciation occurs in the form of ice coatings on the flanks of the highest ridges and as glaciers in steeply-dipping hollows. These glaciers move as blocks separated by deep crevasses owing to the steep slopes, and can form seracs. Ice avalanches frequently originate in this zone. The glaciated peaks rise above the valley of the Rio Santa which is at an altitude of approximately 4000 metres. The snow-line is at an altitude of 4900 to 5000 metres.

Nevado Huascaran, Cordillera Blanca, Peru

8.4 Glacier networks and piedmont glaciers

Glacial networks are said to occur in glaciated areas if the slopes of the glaciated trunk valleys are ice-free up to the mountain peaks. However, valley divides are occasionally completely ice-covered (ice transfluence). The ice-free peaks and ridges between the valley glaciers are called nunataks. As far as areal extent and the relation with relief is concerned, ice networks occupy an intermediate position between valley glaciers (i.e. glaciers confined to fluvially pre-formed valleys) and inland ice masses that completely cover the preglacial relief. The trunk glaciers of the glacier networks merge into a piedmont glacier when they emerge from the mountains. Both ice networks and piedmont glaciers mark glaciation maxima in the Alps during the Pleistocene glacial periods.

Kings Bay, Spitsbergen

8.5 Glacier tor

Water flowing through cracks and crevasses forms subglacial meltwater stream systems in the glacial body. Meltwaters emerge in a stream from under the glacial tongue through a glacier tor (glacier gate; Photo 8.13). This subglacially emerging water is whitish-grey coloured because of its suspended fine material (glacier milk). Glacier tors are a characteristic feature of temperate glaciers, but can also occur on subpolar glaciers during the summer ice melts (Photo 8.6). Glacier tors frequently collapse and the photograph was taken immediately after the ice roof of the tor collapsed, breaking the ice mass into smaller fragments.

Morteratsch Glacier, Upper Engadin, Switzerland

8.6 Glacier calving

The calving of glaciers occurs when large ice blocks (icebergs) are detached along crevasses from the floating glacier tongue or from steep ice fronts. Smaller fragments frequently break away too. The photograph shows the approximately 40-m-high calving front of a subpolar glacier. The subpolar location of the glacier is indicated by meltwater channels on the glacier surface and a glacier tor (centre). Icebergs are floating in the bay, one of which (centre right) shows a clear concave erosional surface formed by wave action.

Blomstrand Glacier, western Spitsbergen

8.7 Glacial 'karst' ('thermokarst')

Stagnant glaciers, if covered by rock fragments and debris, show forms that closely resemble karstic limestone surfaces of doline karsts (Chapter 7). The glacier surface is pitted by small craters or kettles and can also show karren-like forms. Tubes and caves were also created which have since dried up. This convergent range of forms therefore resulted in the term 'glacier karst'. The forms on the surface and in the inner parts of the debris-covered stagnant glaciers are not created by solution processes, but by ablation. Irregularities in the composition of the debris blanket, as well as of the inner ice structure, lead to selective ablation.

Tasman Glacier, Mount Cook National Park, New Zealand (photo by M. Lehr, Tübingen, Germany)

8.8 Dead ice

Dead ice is a stagnant ice remnant that has been separated from a melting glacier or inland ice sheet. The size of dead ice patches can vary widely, and they can either be covered by debris or exposed. The dead ice shown in the photograph is located in front of a 20-km-long and 3-km-wide valley glacier. Cryoconite holes create ablation micro-forms on the surface of the dead ice mass. Dead ice masses can be reactivated by advancing ice sheets.

Mendenhall Glacier near Juneau, Alaska, USA

8.9 Glacial polish and striations
The grinding effects (abrasion) of glacier ice tend to smooth exposed bedrock surfaces (glacial polish), and can even smooth vertical surfaces, as shown here. Embedded debris (right side) can scratch the polished surface, which in this case is a gneiss, and form glacial striations.

Langgrub Valley below the Salurn Ridge, Ötztaler Alps, Southern Tyrol, Italy

8.10 Rock drumlin (*roche moutonnée*)
The grinding effects of glacier ice do not only polish but can also round off sharp-edged rock surfaces. This process forms so-called rock drumlins (or *roches moutonnée*), a typical relief feature, especially in areas previously covered by inland ice masses. Rock drumlins have relatively gradual slopes on their frontal side, i.e. the surface which faces the advancing ice. The photograph shows the rear side that dips in the direction of glacier movement. Plucking has made this surface steeper and less smooth. Following sea-level rise, flooded rock drumlin areas form so-called schären landscapes (Photo 10.13), a typical feature in southern Scandinavia.

near Helsinki, Finland

8.11 Cirque

Initial glacier formation occurs in cirques (Fig. 22). In the example shown here, the moraine-covered cirque threshold or ridge is clearly recognizable. It conceals the cirque basin, which is carved out of the underlying rock and from which the steep backwall rises. Powerful frost weathering processes have formed debris cones on the back and side walls, since the ice melted. When the cirque is ice-filled, the debris becomes embedded and is removed in the glacier. Many cirques are filled by lakes after the ice has melted.

Western flank of the Patteriol, Verwall Group, Vorarlberg, Austria (photograph by Ch. Hannss, Tübingen, Germany)

8.12 U-shaped valley

U-shaped valleys are characterized by steep rocky walls and a rounded basal cross-section. U-shaped valleys have evolved from fluvial V-shaped valleys (Photo 5.1) as a result of glacial erosion processes. In addition to abrasion and plucking, fluvial erosion by subglacial meltwaters also plays a considerable role when temperate and subpolar glaciers erode valleys. The upper limit of glacial erosion on valley slopes is occasionally described as the 'polishing limit'. Fjords are U-shaped valleys that have been flooded by marine transgressions (Photo 10.12).

Roseg Valley, Upper Engadin, Switzerland

8.13 Terminal moraine (recent)

Terminal moraines form walls or elongate mounds in front of glaciers and are created by advancing ice. They consist of material from the medial and especially from the ground moraine, as well as meltwater deposits. They are therefore composed of rock debris, pebbles, gravel, sand and clay. This material is pushed forward by the advancing glacier and piled into a semi-circular wall that marks the ice margin. Older terminal moraines are frequently overridden by renewed glacial advances. The photograph shows the terminal moraine wall of a glacier tongue that is deeply incised by radial crevasses.

Blaisen, Hardangerjokul, Norway (photograph by M. Lehr, Tübingen, Germany)

8.14 Lateral moraine

Lateral moraines consist of rock material deposited along the glacier flanks by melting ice. Alpine glaciers are flanked by 'fresh' lateral moraines from the 1850 glacial maximum. Since then glaciers have been constantly retreating, with the exception of brief periods of advance (Fig. 23). The 1850 moraine is clearly recognizable in the photo, and seracs have formed in a more steeply dipping zone of the glacier, above the glacier tongue.

Tschierva Glacier, Bernina Massif, Upper Engadin, Switzerland

8.15 Ablation moraines

Glacier surfaces can be extensively covered by rock debris (ablation or surface moraine). A prerequisite is strong frost weathering on the slopes above the glacier, after which slope denudation transports the debris on to the glacier surface. The surface moraine material sinks into the glacier as it moves down-valley, forming an inner moraine. This later contributes to the formation of the ablation moraine when ice movements are directed towards the surface in the ablation zone of the glacier. The glacier in this example is completely covered by an ablation moraine which hinders further ablation in the underlying ice. The photo shows the glacier front with a glacier tor and a glacier stream.

Suldenferner, Ortler Massif, Southern Tyrol, Italy

8.16 Terminal moraine (relict form)

The terminal moraine belts in Northern Germany indicate the extent of the Scandinavian inland ice sheets during the Pleistocene. In Eastern Holstein, the 164-metre-high Bungsberg is the highest point of a wall-shaped terminal moraine belt. It was created by the last Pleistocene glaciation and is therefore a relict form. A typical glacial depositional relief, composed of many terminal moraine belts, was deposited along the margin of the Weichselian nordic ice sheet in the area surrounding the southwestern Baltic Sea. This landscape is described as a 'young moraine landscape' owing to the relatively fresh glacial forms that are barely affected by erosional processes. Areas formed during preceding Pleistocene glaciations, which were not covered by the Weichselian glaciation, show glacial forms levelled by Weichselian periglacial erosion. The resulting relief is an 'old moraine landscape' called Geest in Northern Germany.

Area to the southwest of the Bungsberg, Hohwachter Bay, Schleswig-Holstein, Germany

8.17 Knoll-shaped ground moraines

Ground moraines are characteristic features of the hinterland of marginal ice zones (terminal moraine walls) in the terminal basins of glaciers. They consist of rock fragments, glacial till, and glacial clay (a mixture of clay and fine sand). Glacial clay is formed by the grinding up of loose rock fragments during ice movement. The young moraine landscapes of Schleswig Holstein in Northern Germany include large areas of knoll-shaped ground moraines, in addition to the terminal moraine walls. Further to the east, in Mecklenburg and Pommern, level ground moraines predominate. The example shown here is a knoll-shaped or hilly ground moraine.

near Horsens, Jutland, Denmark

8.18 Erratic block

Erratic blocks are large exotic boulders that were transported by ice and deposited at a distance from their original outcrop. All erratic blocks in Northern Germany originated in Scandinavia. As they were transported over long distances, they are usually rounded and frequently scratched by abrading moraine material.

Damestenen, ground moraine on Fünen Island near Hesselager, Denmark

8.19 Dead ice hollow

Level ground moraine surfaces are frequently marked by kettle-shaped hollows with depths of up to several metres and diameters of 20 to 60 metres. They are especially common in Mecklenburg in Northern Germany, and were created by dead ice blocks preserved under moraine material long after the margin of the glacial sheet had retreated. The slow melting of the dead ice remnants led to subsidence of the cover material, thereby forming shallow kettles in the level ground moraine.

near Bad Kleinen, Mecklenburg, Germany

8.20 Pingo

Frost-dynamic processes lead to the formation of pingos, cone-shaped hills that occur in subpolar periglacial areas with permafrost (compare with Fig. 25, including description of their formation).

near Tuktoyaktuk, east of Mackenzie delta, Northwest Territories, Canada (photograph by H. Wilhelmy, Tübingen, Germany)

8.21 Potholes

Potholes are hollow forms abraded into the solid bedrock below a glacier by meltwaters (Photo 2.27). These fluvioglacial erosion forms are created by the grinding effects of rock fragments in rotating turbulent currents below crevasses.The pothole in the photograph belongs to a group of potholes with diameters of up to 6 metres and depths of up to 15 metres.

Maloja Pass, Graubünden, Switzerland

8.22 Esker

Eskers are narrow, elongated, curving wall-shaped ridges of varying heights, which lie parallel to the ice movement directions. Eskers consist of pebbles, gravel and sand and can reach several kilometres in length. They are fluvioglacial sediments deposited in subglacial meltwater channels, in which deposition occurred under frequently-changing hydrostatic pressures. After the ice has melted, they rise above the ground moraine landscape.

near Tammela, southwestern Finland

8.23 Outwash fans (sandur)

Extended meltwater deposits in the form of alluvial fans in glacial forelands are called outwash fans or sandur. The type of material sorting is a characteristic feature of these glaciofluvial accumulations. The inner area of the fan, close to the glacier, dips at a slightly steeper angle and is formed by coarse material, while the outer reaches of the fan are flatter and composed of finer grained alluvium. Repeatedly diverging channels can be observed on glaciofluvial outwash fans, as on alluvial fans (Photo 5.15), and can be seen in the foreground of the photograph. The alluvial fan begins in the middle-ground in front of a Holocene terminal moraine wall. The present glacier tongue, which cannot be seen in the photograph, is further up the valley.

Hooker Valley, Mount Cook National Park, New Zealand (photograph by M. Lehr, Tübingen, Germany)

9 Aeolian landforms

Wind as a geomorphological agent

Wind can only be an effective geomorphological agent if exposed land surfaces are virtually bare of vegetation. This condition is met not only in wide areas of the arid zones, but also in coastal areas under varying climatic conditions. The effectiveness of wind results from its ability to lift material (dust, fine sand and sand, and occasionally also coarse sand and fine gravel) from land surfaces, and to transport and subsequently deposit it in a different location. In addition to accumulation, wind can also cause erosion; its erosional effectiveness is a function of the sediment load (sand blasting).

The ability of wind to lift and transport loose particles is a function of several factors, but predominantly of wind velocities and grain size (Table 4). Increasing wind velocities enable the movement of particles with increasingly larger grain sizes. The transport capabilities of wind are amplified by wind gusting and turbulence which occur more frequently with higher velocities. Only dry loose material can be transported by wind, as moisture binds loose particles by cohesive forces. Additional effects on material transport are exerted by the texture of the basal surface on which the loose material is transported, e.g. by its smoothness and vegetation cover. While sand is mostly transported close to the ground and is therefore strongly affected by surface roughness, dust particles are lifted to greater heights and can then be transported for long distances, for example from the Sahara desert to Central or even Northern Europe.

Sand transport is an especially effective process in creating landforms. After being lifted from the ground by the wind, sand grains move in a series of bouncing steps (saltation) of varying lengths. Saltating sand grains dislodge other grains when they strike the surface again, and these are then caught and transported by the wind. Larger grains (coarse sand and fine gravel) that cannot be lifted by the wind, can be pushed (creep) by the percussive force of saltating fine grains. These processes lead to continual sand redistribution, and the subsequent creation and destruction of landforms.

Table 4 Wind velocities and transported grain sizes (from Leser & Panzer 1981)

Fraction	Grain diameter (mm)	Wind velocity (m/s)
Dust	to 0.01	0.1
Fine sand	to 0.03	0.25
	to 0.04	0.5
	to 0.1	1.0–1.5
	to 0.12	1.6–1.7
Medium sand	to 0.25	1.8–3.3
	to 0.4	3.4–5.2
	to 0.5	5.3–7.4
Coarse sand	to 0.75	7.5–9.8
	to 1.0	9.9–12.4

Aeolian erosional processes and forms

Aeolian erosion can be subdivided into two main processes: an evacuation of material (deflation), and the abrasive effects of wind-carried material (wind corrasion) which is similar to sand-blasting. Deflation can form basin-shaped hollows (deflation hollows) on a micro- or macro-relief scale. They occur frequently in desert areas where less-resistant rocks are exposed. Deflation effects, partially in conjunction with fluvial or surface washing, are also indicated by specific meso-relief forms in arid zones (Chapter 5), e.g. hamadas (Photos 5.32, 5.33) and serirs (Photo 5.40). All of the fine-grained disintegrated rock material generated by weathering in the hamadas, as well as the loose material accumulated by fluvial or slope wash in wadis or serirs, is evacuated. This creates exposed or debris-covered rock or pebble surfaces (stony desert or gibber plain). Evacuated rock pavements (deflation pavements) are not restricted to desert areas but can also be recognized as relict forms in cool-temperate climatic zones in areas of Pleistocene sandy glacial deposits, where they were formed under arid cold climatic conditions. They often contain windkanter or dreikanter, that is, rock fragments that have been polished by sand-blasting. Predominant, constant wind directions can create surfaces or facets on these fragments, and several smooth facets that meet in sharp edges can indicate either changing positions of the fragment, or dominant winds from different directions, e.g. bidirectional winds. In addition to these micro-forms, varying degrees of rock resistance (Chapter 2) can form natural arches (Photo 2.15), concave corrasion hollows and mushroom-shaped rocks (Photos 2.14, 9.1).

Aeolian depositional processes and forms

Aeolian sand deposits occur in many forms. Blowing sand can be deposited in aerodynamically-shaped sand drifts (Photo 9.2) on the sheltered side of obstacles. Micro-forms, such as small accumulations (Photo 9.3), can develop around the base of plants when loose sediments are held by roots, and windblown sand is effectively trapped by the plants. The meso- and macro-forms of aeolian deposition, sand seas (or sand sheets) and sand dunes, are much more impressive than the micro-forms. Sand seas form areas of extended monotonous relief, while dunes occur in a wide variety of shapes and sizes. In sand seas (Photo 9.4) more than 40 per cent of the accumulated sediments consist of coarse-grained material (coarse sand and gravel), while fine sand contents in dunes typically exceed 80 per cent. The relatively high percentage of coarse-grained material in sand seas hinders dune formation, and ripple marks (Photo 9.7) also only play a subordinate role. Ripple marks are small, subparallel, ridge-shaped sand accumulations. They are formed by wind action at the boundary layer, in this case between air and sand, but are basically similar to those that occur on beaches between water and sand. Ripples trend at right angles to the direction of air or water movements. Their wave length is in the order of centimetres and is a function of wind velocity and grain size.

Dune areas are composed of separate sand accumulations in the form of individual meso-forms that in turn form macro-relief areas of considerable extent (erg). Fine-grained sand sizes are a typical feature of all dune-covered desert areas. Areas in which dunes occur can be subdivided into desert, coastal and inland areas. An additional criterion for dune classifications is their shape. The main differences are between fixed and moving sand dunes, and dunes that are positioned at right angles or parallel to the prevailing wind direction.

Fixed dunes are sand drifts created against or behind obstacles such as vegetation or relief features. The resulting micro-forms on the protected side of obstacles have already been described. Large 'lee-side' fixed dunes (Photo 9.5) are mostly formed by specific relief features such as large obstacles. Even though their size increases, these dunes remain immobile as they cannot become detached from the obstacle that created them.

Moving dunes are a much more common feature in desert areas than fixed dunes. A typical moving dune is the barchan (crescentic dune; Fig. 26a, Photo 9.6) in which the entire sand mass is moved. Barchans occur as isolated sand bodies on rocky or gravelly substrates. Other types of moving dunes in desert areas are represented by long, narrow transverse or longitudinal dunes, and by aklé dunes which are sub-units of complex transverse dunes formed by coalescing barchans. The aklé dunes occur in subtropical deserts, where seasonally-controlled bidirectional winds blow in opposite directions (trade-winds in summer and west winds in winter). The corresponding sculpting processes on the exposed and protected flanks are seasonally alternating, and the curved shapes of a chain of aklé dunes therefore also change their direction biannually. In addition to transverse dunes, the central areas of the giant dune fields (ergs) also contain very high (100 metres) and several-kilometre-long sand dunes which point in the direction of the wind. These giant dunes, called draas, have a basal width of 1 km and form parallel chains about 2 to 3 kilometres apart. Their origin is difficult to determine, but their development is now explained by the presence of parallel turbulent flow rolls. These are oriented with the wind and adjacent turbulent rolls are always contra-rotating. Longitudinal dunes in desert areas also include seif dunes (sword dunes) and silk dunes (line dunes; Photo 9.7). While all of the above-mentioned dune types can be classified as primary dunes, secondary dunes can also be recognized as separate structures on larger, older dunes. The best-known type of secondary dune is the star dune (pyramidal dune or ghourd; photo 9.8).

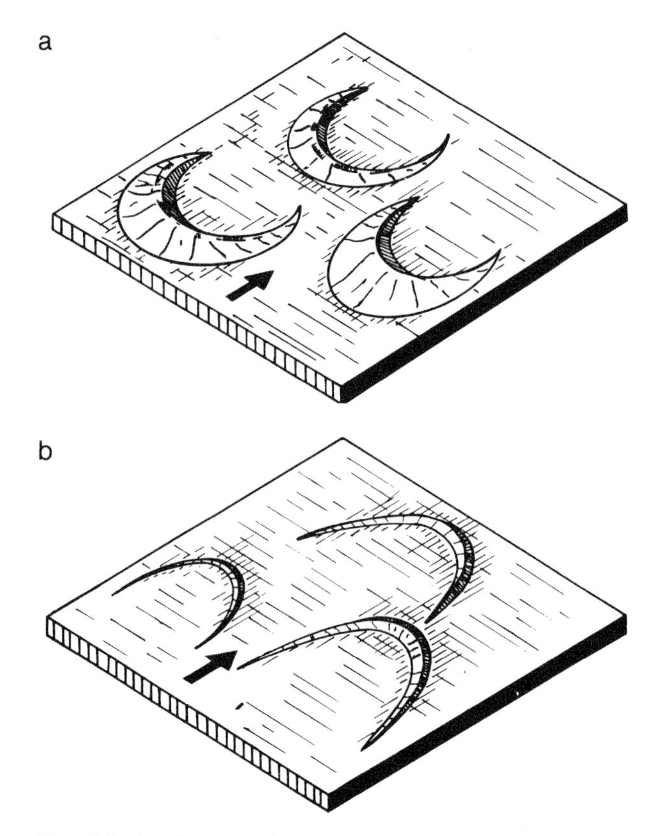

Figs. 26a, b Crescent-shaped isolated dunes

a Barchan (crescentic dune) in deserts (Photo 9.6)
b Parabolic dune in coastal and inland areas

The arrow indicates wind directions. In barchans the crescent horns precede the central part of the dune, while in parabolic dunes the central part moves more rapidly than the horns.

Parabolic dunes (Fig. 26b) are characteristic developments in coastal and inland areas. They are mostly recent features in coastal environments but relict forms in inland areas, and can, for example, be found frequently in the humid cool-temperate zones of Central Europe, where they were formed under periglacial arid-cold climatic conditions during the Pleistocene. Parabolic dunes resemble barchans, as they also occur as isolated crescent-shaped features, but they are reversed, that is the horns point in the upwind rather than downwind direction. The central part is steeply banked on the wind-protected outer flank of the crescent, whereas barchans have steeper slopes on the lee-side inner flanks. The concave opening of parabolic dunes therefore faces the wind and this is the primary differentiating feature from barchans. In coastal areas, ground moisture and vegetation restrict sand movement, especially in the horns of parabolic dunes. Despite its larger mass, the central part of the dune migrates more rapidly than the horns. Wall dunes (fore dunes; Photo 9.9) at right angles to formative wind directions are also frequent occurrences in coastal zones.

Although dune forms are indicative of constant sand transport and resedimentation processes, the areal extents of the great desert dune fields (ergs) are comparatively well defined. They are limited to relative lows (basins). It was previously felt that erg areas, with their characteristic sandy features created by aeolian accumulation processes, represent typical desert areas. However, aeolian accumulation forms only cover a relatively small percentage (14–20 per cent) of the arid zones, even though individual erg areas can extend over several 100 000 km². Sand that accumulates in ergs is partially derived from surrounding dry valleys, but a more important source is weathering products from adjacent pediments and serirs. However, the highest percentage is provided by weathering on exposed rock surfaces (hamadas), especially on sandstones. Erg sands are therefore mostly weathering products from earlier phases and have experienced repeated reworking by erosion and sedimentation. A general rule is that large dune areas have polygenetic histories.

Dust deposits

Dust evacuation and deposition can also occur in non-arid zones, for example during dry periods in humid climatic zones, especially if the natural vegetation cover has been removed by anthropogenic influences. However, special forms do not result from these processes, with the exception of the deposition of loess which reaches thicknesses of up to 40 m in Central Europe and of more than 400 m in Eastern Asia. Loess deposits blanket slopes to varying thicknesses and can also cover, as for example in China, drainage divides. Loess is calcareous dust deposited by aeolian processes, and consists mainly of quartz particles with grain sizes of less than 0 · 06 mm (silt) together with 8–20 per cent calcareous material. Recent evacuation of loess is, for example, occurring in the cold desert areas of Asia. A prerequisite for its deposition in the adjacent steppe areas of China is the presence of a light vegetation cover that traps the silty dust. Most of the world's loess accumulations are prehistoric as they were deposited under arid-cold climatic conditions during the Pleistocene. This is also the case for the loess deposits of Central Europe, where evacuation occurred from moraines and sandurs, from gravel floors in valleys, and from debris fans.

9.1 Formation by wind corrasion (mushroom-shaped rock)

The development of overhanging cliffs and mushroom-shaped or pedestal rocks is usually caused by resistance differences between adjacent rock layers. The ubiquity of these forms is related to the fact that selective weathering and slope denudation processes are active under all climatic conditions, even if the actual processes are different. The mushroom-shaped rock shown here is located in the Eastern Sahara in an extremely arid zone with annual mean rainfall values of less then 3 mm. Insolation weathering, gravitational forces and basal wind corrasion are the erosive processes that have led to its formation. Water washing processes at the base are the only processes that have not been active in this case, unlike the example shown in Photo 2.14.

near Wadi Halfa, Sudan (photo by H. Wilhelmy, Tübingen, Germany)

9.2 Aeolian accumulative micro-forms: sand drifts

Accumulations of windblown sand form on the wind-protected side (lee-side) of many obstacles. In the example shown, small 2–3-cm-long sand mounds have been formed on the lee-side of shell fragments by winds from the east. A scale is provided by a 9 cm long shell (*Ensis arcuatas*) in the front right hand corner of the photograph. When the wind changed to the west on the day after the photograph was taken, the newly-formed tiny mounds all pointed in the opposite direction. Larger obstacles lead to correspondingly higher, longer and wider aerodynamically-formed sand drifts.

Beach surface at St Peter-Ording, Eiderstedt, Schleswig-Holstein, Germany

9.3 Nebkhas

Small mounds, for which the Arabic word nebkhas is occasionally used, can be formed by deflation or water wash processes, or by aeolian accumulation. The roots of grass, bushes, or trees protect the loose sediments from being removed (Photo 9.9) or provide traps in which sand accumulates. The nebkhas in the example are located on the margin of a dry valley. They are erosional forms, as indicated in the foreground. In the middleground sand accumulations are forming, and (behind the footprints) initial dune formation can be observed. Nebkhas can reach a height of 20 m if the enclosed plants manage to grow as quickly as the sand accumulation.

Foreland of the Sahara-Atlas east of Biskra, Algeria

9.4 Sand sea (sand sheets)

Sand seas are characterized by very low relief, even if they cover extensive areas of several 100 km^2. A high percentage of the accumulated material consists of coarse-grained fragments that hinder dune formation. Ripple marks (foreground) are rare and only partially formed.

near Umm al Araneb, Fezzan, Central Sahara, Libya

9.5 Lee-side dunes

Lee-side dunes are fixed dunes attached to a specific obstacle. Smaller fixed dunes are mostly attached to plants, but larger lee-side dunes can be relief-controlled. The formation of the lee-side dune in the example shown here can be attributed to the presence of a dome-shaped rocky outcrop, against which the wind has blown sand from a wide, mostly dry valley floor (from right to left in the photograph). Sand has been deposited on the wind-protected lee-side, and windblown sand, which has formed the dune, can be seen in the right background. Less than 30 mm annual mean rainfall characterizes the desert climate in this area. Agricultural use is dependent on irrigation and is restricted to areas near the mouths of rivers that flow from the Andes. These are flooded during February and March and transport enormous debris loads. However, during most of the year, the valleys are dry and abundant material is available for aeolian removal. In this case, regular strong winds blow down-valley during the night until late morning and maintain this dune on the lee-side of the rocky dome where the valley merges into the coastal plain.

near Casma, northern coastal desert, Peru

9.6 Barchan (crescentic dune, Fig. 26a)

Barchans are isolated dunes in which the entire sand mass is moved. Their characteristic feature is the crescent-shaped form with drawn-out crescent ends (horns) that point downwind (leeward). The slope on the windward side has a shallow angle, while on the lee-side it is steeply angled (30°). Wind turbulence on the lee-side of the barchan clears the ground between the dunes. The central part of the barchan is a massive, high sand body while the crescent ends are narrow and low. These move more rapidly than the central part as their mass is lower. Barchans can be up to 20 m high and a typical 5-m-high dune can move at a rate of 20 m per year. An increase in height of 2 m reduces the annual migration rate by about 2 · 5 m. As indicated on the photograph, barchans mostly form on hard ground.

near Santa Isabel de Sihuas, southern coastal desert, Peru (photograph by H. Wilhelmy, Tübingen, Germany)

9.7 Silk dunes (line dunes)
The Arabic term 'silk dune' is often applied to low, closely-spaced dunes with curved crests. They are generated by the lateral linking of seif dunes (sword dunes). Seif dunes are low, curved wall-shaped dunes with similar slope angles on both the windward and lee-sides. The sinusoid crestal shapes of silk dunes are a result of their origin as seif dunes that have grown together to form longitudinal dunes aligned to the prevailing wind direction. A prerequisite for their formation is the existence of seasonally-changing, bidirectional wind systems. In the dune field shown here, the northwestern shamal winds change to east winds in spring. Ripple marks can be seen on the dune surface in the foreground.

near Hofuf, Eastern Province, Saudi Arabia

9.8 Star dune (pyramidal dune, ghourd)
Star dunes are the most common type of secondary dune, and most often occur in large dune fields (ergs). The characteristic feature of these sand accumulations is their star-shaped structure and their position on top of larger, older dunes. Star dune heights can exceed 100 m. In the example shown the sinusoidal crestal shapes are silk dunes.

Foreland of the Adrar Acacus near Ghat, Central Sahara, Libya

9.9 Fore dunes
Wall-shaped dunes (fore dunes) mostly form in coastal environments. They are positioned at right angles to the formative wind direction and are therefore classified as transverse dunes. Wall dunes are wide ridges with gentle windward and steeper lee-side slopes. They migrate down-wind until stopped by vegetation. The wall-shaped moving dune in the photo has strong surficial deflation marks such as hollows and channels. The roots of the isolated grass clumps prevent deflation, so that nebkhas (Photo 9.3) are formed as erosional remnants. These remain until the plants die.

west of List, Island of Sylt, Schleswig-Holstein, Germany

10 Marine landforms

Marine forming processes: abrasion and accumulation

The rise and fall of oceanic water levels means that the margin between the land and water surfaces of the earth is not a sharp line. It is a marginal zone where processes of erosion and accumulation create a variety of coastal forms. The rhythmic tidal rise and fall of water levels occurs in 6¼-hour cycles. The difference in height between high and low water levels, the tidal range, is usually less than 2 metres, but can be much higher along some coastlines, reaching maximum values of 21 metres. Long-term sea-level fluctuations are of eustatic origin, i.e. they are caused by climatically-controlled changes of oceanic water contents. For example, during the Pleistocene glacial periods, sea-levels fell considerably (marine regression) as large amounts of water were trapped in the extensive glaciated areas. During the last ice age, sea levels were approximately 100 m lower than at present, while in the warmer periods, sea levels rose again (marine transgression). Tectonic uplift or subsidence of coastal areas also causes marine transgressions or regressions, although these are mostly only local.

Wave action is the most important process and can cause both abrasion and accumulation. Many coastlines have alternating sections of erosion and accumulation. The sea can act as an abrasive force on protruding parts of the coast, while often accumulating material in bays. If the outline of an originally irregular or indented coastline has been smoothed by erosional and accumulation processes it is often described as a graded shoreline.

Abrasion

Wave action on steep coastlines (see below) creates most impressive erosional morphologies. Typical abrasive forms are shore platforms (Fig. 27, Photo 10.1), concave abrasion notches (Photos 10.1, 10.2), wave-cut clefts (Photo 10.3), sea caves (Photo 10.4), natural or sea arches (Photos 10.1, 10.4), and resistant residuals or stacks (Photo 10.1).

Accumulation

Wave movement also leads to sediment accumulation, especially in low-relief coastal areas. Waves that strike the coast at an oblique angle and longshore currents move material along the coast (longshore drift; Photo 10.5). Elongated sand accumulations (sand bars) become attached to protruding coastal sections to form spits pointing in the direction of longshore currents. Spits can close off a bay, resulting in barrier beaches (*nehrung*, Photo 10.6) that can enclose a lagoon. Waves that approach at right angles to the coastline build barrier beaches (Photo 10.7) at the mean high tide level, while offshore bars (Fig. 28) are established parallel to the coastline. These ephemeral, fluctuating features are formed between the zone revealed at low tide and the zone just below the low tide level. On the margins of oceans with warm surface waters, i.e. in the tropics and subtropics, so-called beach rock (Photo 10.8) can form from sandy barrier beaches.

Coastal forms

Several factors determine whether abrasive or accumulative processes are dominant along coastlines and which resulting forms then characterize the coast. A

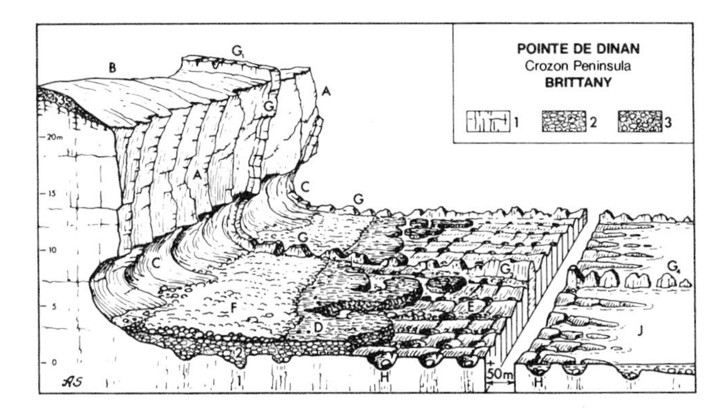

Fig. 27 Cliff coastline, Crozon Peninsula, Brittany, France (from Schou 1968)

A: Cliff. The detailed form of the cliff face results from differential rock resistance and fracturing. Ordovician shales (1) are exposed in the cliff and on the abrasion (shore) platform that extends in front of the cliff.
B: Land surface, which caps the shales as a sculptured surface and is covered with solifluction debris (3).
C: Wave abrasion notch.
D: Abrasion platform in older beach gravels (2) currently being removed by abrasion.
E: Recent abrasion platform. Ridges and channels of the micro-relief are caused by different rock resistances and fracturing.
F: Recent barrier beach.
G: Resistant quartz vein.
H: Pothole.
I: Low water level.

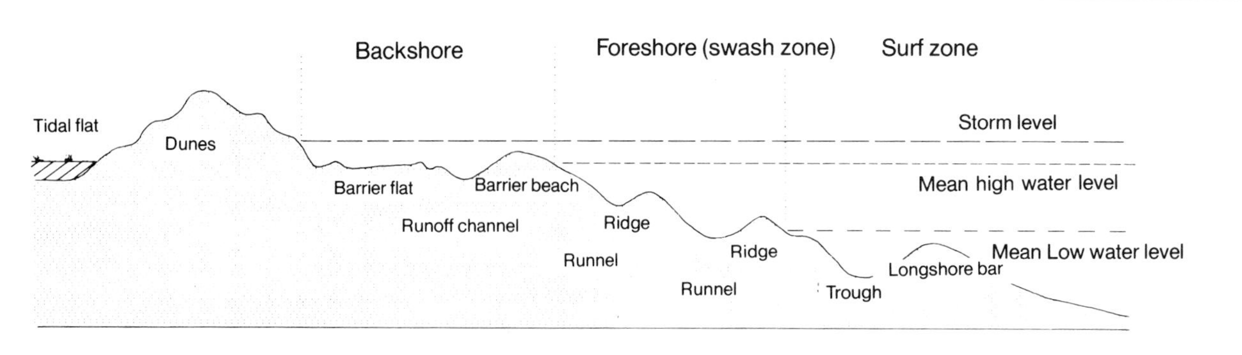

Fig. 27 Barrier beach island exemplified by an East Frisian island, schematic cross-section (from Hempel 1985)

decisive factor for the formation of coastal relief is whether marine transgressions or regressions are occurring. Other factors which determine coastal forms are geological structure, lithological types, and landforms of the onshore coastal zone created by subaerial erosional processes. Finally, coastal relief is also controlled by climatic conditions. Classifications of coastal forms have been performed using each of these criteria individually. However, the following subdivision takes several of the factors that affect coastal forms into account.

Receding coastlines

Coastlines recede during marine transgressions. The resulting submerged coasts can show many different forms, as the flooded areas are subaerially pre-formed. Examples of submerged coast forms in areas previously shaped by fluvial denudation are ria (Photo 10.9), vallone (Photo 10.10), and bolsa coasts (Photo 10.11). Examples of submerged glacially pre-formed relief are fjords (Photo 10.12), schären (or skerries in the Orkneys) (Photo 10.13), and förden coastlines (Photo 10.14). In all of these examples of inlet formation, marine processes only play a relatively insignificant role. However, marine processes are much more intensive along coastal sections bordering open seas, as these are exposed to the direct effects of wave action. The typical erosional forms of abrasion – cliffs – are created which also indicate receding coastlines. The widely varying shapes of these destructional forms is mainly a function of the landforms of the respective coastal area, as well as of the resistance of the rocks

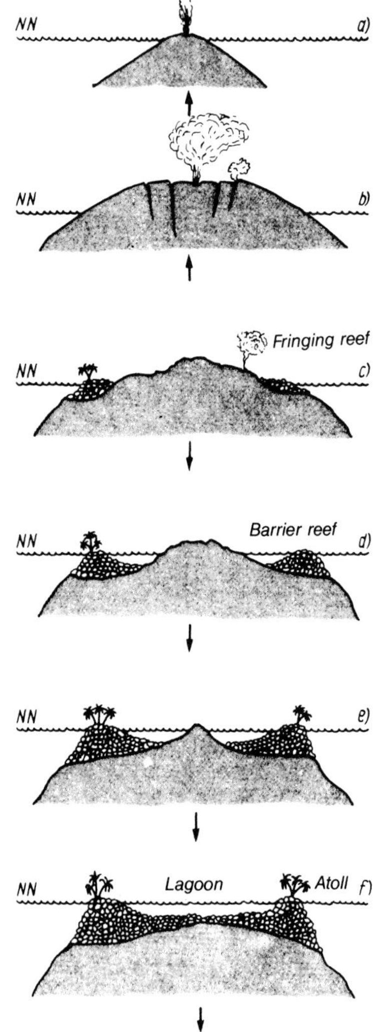

Figs. 29a–f Atoll development stages (from Rast 1987)

a A submarine volcano emerges from the sea.
b Growth of volcanic island.
c Large lava accumulations start to sink, reduced volcanic activity. Initial abrasion, cliff formation, and slope denudation. Fringing reefs are formed by corals at island margins.
d Subsequent sinking leads to barrier reef formation.
e Initial atoll formation. Only an erosional remnant of the former volcanic island remains.
f Atoll stage. The volcanic island has subsided further and has been completely covered by coral structures. The coral reefs completely surround the lagoon.

that form the cliffs (Photos 10.15–10.18). In all cases, the processes of slope denudation (Chapter 2) play a fundamental role in cliff formation, in addition to marine abrasion (Photos 10.16, 10.17).

Advancing coastlines

Advancing or emergent coastlines develop when the border between land and sea moves towards the sea. This applies to coastal areas which emerge during marine regressions as well as to built-up coasts created by accumulation or by organogenic structures. Emerging coastal areas can show, for example, abrasional terraces (Photo 10.19), and abandoned cliffs (Photo 10.20). Built-up coastlines include deltas (Photo 5.17), which are created by fluvial accumulation and form more or less triangular sedimentary bodies in the sea. On the other hand, coastal plains (Photo 10.20) are marine accumulations which stretch along coastlines and frequently intrude into bays. Their seaward side is mostly formed by a barrier beach. Island chains such as cays or keys (Photo 10.24), which only occur in shallow tropical seas, also have built-up coasts. A globally occurring feature along low-relief coastlines with high tidal ranges are tidal flats (Photos 10.21, 10.22). They are also formed by accumulation processes and are partly controlled by plant growth which traps mud. This process is especially effective in mudflats in tropical seas (Photo 10.23) where mangroves occur. Predominantly organogenic coastal accumulation is represented by coral structures that are restricted to seas with warm surface waters. Water temperatures between 25 and 30°C offer optimum conditions for reef-building corals. They cannot survive if temperatures are always below 20°C and their preferred environment reaches from just below the low water level and, as they require light, down to a water depth of about 40 m. Coral structures can be differentiated into fringing reefs (Fig. 29c), barrier reefs (Fig. 29d), and atolls (Fig. 29f, Photo 10.25).

10.1 Abrasion platform, cliff, sea arch, stack

Abrasion platforms (Fig. 27) are gently sloping surfaces that extend seaward from the base of cliffs and are formed by abrasion (erosion resulting from wave action). Much of the platform area is below the mean high water level and is therefore exposed at low tide. Abrasion platforms are rock surfaces on which rock material from the cliff and from the platform itself exerts a grinding effect when moved across it by the sea. Different resistances of outcropping rocks result in selective erosion and the creation of microforms. A barrier beach can form on the uppermost part of the abrasion platform, as shown in the example. Corrasion by rock fragments and air pressure from wave action abrade the cliff face by loosening rock material and then by plucking fragments from the surface. Chemical corrosion (Photos 7.23, 7.24) also contributes to the erosive forces on cliffs. All of these selective erosional processes lead to the development of specific forms on the cliff face: erosional notches (Photo 10.2), sea caves (Photo 10.4), and stacks (left hand side of photograph). The abrasion platform and the 70-m-high cliff are formed in soft Cretaceous limestones.

near Étretat, Pay de Caux, France (photo by Ch. Iven, Rösrath, Germany)

10.2 Abrasion notches

Notches (concave abrasional notches, Fig. 27) can be observed along the upper end of abrasion platforms, i.e. at the base of cliffs near the mean high water level. They are created by the processes mentioned in the description of the previous photograph. The coral limestone rock shown here is completely surrounded by a deep abrasional notch. The isolated erosional limestone remnant is rapidly being destroyed by the joint effects of wave corrasion and limestone solution processes, as indicated, for example, by the holes in the surfaces.

Ile aux Bénitiers, southwest coast of Mauritius

10.3 Wave-cut clefts

Corrasion forces can penetrate deep into solid rocks along fractures that form zones of weakness. Surf action extends these fractures to form abrasional clefts. Resistance differences between adjacent layers and the existence of bedding planes can also amplify the effectiveness of surf corrasion. The eroded clefts in the example shown here have developed in fractured Paleozoic sandstones which form a 60–70-m-high peninsula.

Cape Fréhel, Gulf of St Malo, Brittany, France

10.4 Notches and sea arch

Fracturing can favour geomorphological processes which are active in the surf zone, as listed in 10.1, and can form sequences of abrasional sea caves. The sea caves shown here are mostly related to fractures in limestones. At low tide, the caves and the sea arch (right foreground) are exposed.

Pointe de Dinan, Crozon Peninsula, Brittany, France

10.5 Longshore drift

Longshore drift transports sand and gravel along coastlines, partially due to longshore currents parallel to the coast, and partially to waves that strike the coast at an oblique angle. Longshore drift follows the direction of the prevailing winds. In the example shown here, where the base of the Cretaceous limestone (chalk) cliffs is protected by concrete from abrasion, the drift direction is away from the viewer, and the drift processes can be recognized by the deposits between the groynes.

Channel coast east of Brighton, England

10.6 Barrier beach (*nehrung*)

At their most extreme form, barrier beaches occur as narrow peninsulas which can completely close off a bay area. They can develop when sediment accumulations, which are initiated at coastal protuberances (spit formation), extend over to the opposite side of a bay. They can also be formed by two spits growing together from opposite directions, or by an offshore bar being pushed onshore. A bay that is partly or completely separated from the sea by such a barrier beach is called a lagoon. The barrier beach in the example connects the British mainland with the Isle of Portland. It does not consist of sand, but of gravel - mostly flints from the Cretaceous limestone (chalk). In the background, separated from the sea by a man-made dam, a lagoon with brackish water can be seen. The gently curving barrier beach is 29 km long, and its width increases gradually with increasing distance from the mainland to reach 183 m at the Isle of Portland. The height of the barrier beach increases in the same direction to reach a maximum of 13 m above the mean high water level (middleground left). A harbour that has been protected from the open sea by man-made structures since the middle of the nineteenth century is also located in this area.

Chesil Bank, view from the Isle of Portland, English Channel coast (photograph by Ch. Iven, Rösrath, Germany)

10.7 Barrier beach

In the backshore zone along the coastline, wave action builds up barrier beaches which run parallel to the ridges in the foreshore (swash zone) and surf zone (Fig. 28). The crests of these flat accumulations consist of sand or gravel, and mark the maximum reach of waves at high tide. The barrier beach in the example connects the island of Margarita to the small adjacent island of Macanao. The left-hand side dips gently into the open Caribbean Sea, while the right hand (southern) side merges into a mangrove lagoon (not visible in the photograph). On the seaward side, the barrier beach shows concave indentations caused by wave action.

Istmo Restinga, Island of Margarita, Venezuela

10.8 Beach rock

On the margins of oceans with warm surface waters, i.e. in the tropics and subtropics, beach rock can form on barrier beaches if these consist predominantly of calcareous sand. Beach rock thicknesses can range from several decimetres to more than 2 m. Beach rock is a calcareous sandstone crust which forms near ground-water level by calcium carbonate precipitation from sea and/or fresh water. Sufficient quantities of calcium carbonate in sea water are a precondition for sand grain cementation processes, such as those which occur to form beach rock. Calcium carbonate is obtained from algae, corals, and shells. The beach rock in the photograph is being destroyed by abrasion, as indicated by the platy shapes. Similar forms are created on barrier beaches by the precipitation of calcium carbonate from surf spray.

Heron Island, southern Great Barrier Reef, Australia (photograph by Ch. Iven, Rösrath, Germany)

10.9 Ria coast

Rias are inlets or arms of the sea created by the drowning of a valley (monofluvial) or of several valleys (polyfluvial) during sea-level rises. The name is from the type location in Galicia (Spain). The surrounding relief is one of low fossil peneplains, which were fluvially incised to create valleys before the sea entered the lower reaches of the valleys. Monofluvial rias are long, narrow arms that reach deep into inland areas up to a distance of several tens of kilometres. The example shown here is polyfluvial and forms an extremely complex system of bays and inlets and more than fifty islands, with a maximum width of 22 km, a surface area of 100 km², and a maximum water depth of 15 m. Many inlets are dry at low tide. Together with the western monofluvial ria of the Rivière d'Auray, they form a gulf connected to the open ocean by a 900-m-wide mouth.

Gulf of Morbihan, south coast of Brittany, France

10.10 Vallone coast

A vallone coast is formed by the drowning of longitudinal valleys in mountain chains. Similar to monofluvial rias, they are long, narrow arms of the sea. However, their lateral slopes are much steeper and higher. The vallone shown here is 10 km long. If these inlets run parallel to the coastline, they are called canali.

Limski Zaliv, north of Rovinj, Istria, Yugoslavia

10.11 Bolsa coast
Wide, bottle-necked bays with narrow openings to the sea are called bolsa bays from the Spanish word for a 'large purse' used for these forms on Cuba. They indicate submerged coasts along which the narrow inlets represent drowned gorges through mountain chains or rocky barriers. One of these – a seaward-dipping bank of coral limestones – can be seen in the background. The bay widens inland into an area of fluvial denudation behind the limestone barrier.

Sint Marie Bay, Curaçao, Netherlands Antilles

10.12 Fjord
The most impressive coastal forms in glacial areas are fjords. They are U-shaped valleys (Photo 8.12) into which the sea has penetrated. Characteristic features are steep walls, narrow width, and great length. The strength of the glacial erosion processes, a typical feature of glaciated mountain areas, is indicated by the considerable depth of these inlets. The Sogne Fjord in Norway, for example, reaches a depth of 1300 m. As a result of typical glacial erosion processes, the deepest part of the fjord is not at its mouth.

Jössingfjord near Egersund, Norway

10.13 Schären
Schären (also called skerries in the Orkneys) are small, rocky islands that occur in great numbers where low-lying areas of glacial erosion have been flooded by the sea. Many of the small hills formed by ice movements (*roche moutonnée*, Photo 8.10), rise above sea-level as schären islands.

Skagerrak coast near Strömstad, Sweden

10.14 Förde

Elongate bays in glacial depositional areas are described as förden from their type locations in Northern Germany and Denmark. They are formed when glacial tongue basins or subglacial meltwater channels are drowned by rising water levels. The förde shown here is a drowned glacial tongue basin, formed during the last Pleistocene glacial period, with varying widths of up to 6 km, depths of up to 10 m, and a total length of 16 km.

Kieler Förde, view from Laboe, Germany

10.15 Cliff in resistant rocks

Erosion by wave action proceeds more slowly on cliffs of resistant rock types than on less resistant material. The cliff shown here, with a height of up to 200 m, consists of a base of resistant Carboniferous limestones and is fully exposed to the waves of the open Atlantic Ocean. Spray from the breaking waves can strike the cliffs at heights of up to 100 m during storms. The upper part of the cliff is composed of alternating, thin limestone and shale layers and the overall slow retreat of the cliff face is mainly the result of the recession processes described in Chapter 2. Resistant limestone layers are selectively weathered out as ledges by these processes.

Cliffs of Moher, County Clare, Ireland

10.16 Cliff in less resistant rocks
Erosion and cliff recession is a more active process in less resistant rocks. The cliff shown here consists of soft Cretaceous limestones. Strong abrasional processes are active at the base of the cliff, and the steep upper section is the product of slope denudation, indicated here by massive gravitational rock slides. The loose material created by abrasion and slope denudation is then removed by wave action.

English Channel, southwest of Calais, France

10.17 Cliff in loose material, strong erosion
If cliffs composed of unconsolidated material are exposed to the open sea, coastal recession can be extremely rapid. For example, terminal moraines deposited during the last ice age in Northern Germany are being aggressively eroded by extreme high tides during winter storms. Intensive slip and glide processes in the water-saturated material contribute to the recession, which reached 60–70 m along exposed north-eastern coastal sections over the period 1875–1950, i.e. rates of 0·8 m per year (near Travemünde). In the section shown here, rates of 0·3 m per year were attained. The footpath along the top of the cliff must therefore be moved back every few years.

near Danish Neinhof, north of Kiel, Schleswig-Holstein, Germany

10.18 Cliff in loose material, moderate erosion
Cliffs in unconsolidated material can also be eroded by abrasion and slope denudation in protected bays, although this is usually less extensive than on open coasts. The boulders shown in the photograph, which form a barrier along the base of the cliff, are from a ground moraine that cut into the cliff during the last ice age.

near Clifden, County Galway, Ireland

10.19 Abrasional terraces
Abrasional terraces are indicative of emerging coastlines. Coastal advance is due either to tectonic uplift of the mainland or to eustatic sea level lowering. In a process similar to river terrace formation, terraces have developed at various levels to form a terrace flight. Remnants of an abrasional terrace can be seen in the fore- and middleground of the photograph. The terrace has only been partially preserved, as abrasion and slope denudation during later, lower sea-level stages have partially destroyed it.

Pacific Coast, north of San Francisco, California, USA

10.20 Abandoned cliff

Abandoned cliffs are in a sense similar to abrasional terraces, as they also indicate coastal emergence. They are called abandoned as they are no longer subjected to abrasion, often because of marine regression. In the case illustrated here the change from active abrasion of the cliff to inactivity is associated with the formation of a coastal plain. The plain is an inorganic accumulation, i.e. without a contribution from plants or animals, that has been created by marine processes, and a barrier beach borders the sea. Many such plains have formed on the shores of the Mediterranean since the days of antiquity, so that old harbours are often no longer located on the coast. Even under natural conditions, rivers in the subtropical winter-humid Mediterranean areas carry considerable amounts of suspended material, sand, and gravel (Chapter 5). Material transport has been intensified by large-scale deforestation. This caused rapid expansion of the alluvial fans of rivers, with subsequent intensified longshore drift and the growth of accumulative coastal plains.

Mediterranean coast near Calahonda, Andalucia, Spain

10.21 Tidal flats

Tidal flats, a specific form of built-up coastline, can develop along low-lying coasts with tidal movements if abundant supplies of unconsolidated sediments are available. High tides cover the entire area which is then exposed during low tide. Tidal flats, such as those on the North Sea coast of Germany, can be very extensive. They form a wide band along the coastline. The deeper outer tidal flat areas are sandy, while the innermost areas close to the shoreline consist of muds. Fine-grained material is deposited in the latter environment during the transition from high to low tide. Mud sedimentation is enhanced by halophytic plants (e.g. *Salicornia herbacea*) in the uppermost tidal flat zones, up to 20 cm below the mean high water level. Tidal flats provide an ideal environment for a rich fauna of mussels, snails, crabs, and worms. As shown in the photograph, abundant fresh excrement heaps of the sandworm (*Arenicola marina*) cover the surface of the sandy mudflats.

near Munkmarsch, Island of Sylt, Schleswig-Holstein, Germany

10.22 Tidal creek

Tidal creeks are small channels in tidal flats and salt marshes in which the water-flow direction changes according to the tidal stage, i.e. whether water is flooding or draining from the flat. They reach above the mean high water level, which marks the boundaries of the tidal flats, into the salt marshes – an area covered with grassy halophytic plants. These salt grass areas, into which the creeks are incised, are only flooded during occasional extremely high tides. On the ebbing tide, water drains from the flats through the complex network of tidal creeks, leading to strong erosion. The main creeks, or tidal channels, can therefore reach considerable depths of up to several metres.

near St Peter-Ording, Eiderstedt, Schleswig-Holstein, Germany

10.23 Mangrove coast

Mangrove coasts are limited to the tropical zones. They are a special form of tidal flat, and the characteristic accumulation process is amplified by the presence of mangrove vegetation. Mangroves can only survive when their environment is protected from direct wave action by a coral reef or a barrier beach. They therefore mostly occur in lagoons. Effective mud trapping is performed by the roots of halophytic trees. *Rhizophora* has curved, stilt-like roots, while *Sonneratia* and *Avicennia* can be identified by a multitude of vertical, asparagus-shaped air roots (pneumatophores, foreground).

north of Tulear, southwest coast of Madagascar

10.24 Cay (Key)

Chains or rows of small islands are typical features in shallow tropical seas. They are called cays after their type locality in the Bahamas (keys in Florida). The Bahama Cays and the Florida Keys are islands on the slightly elevated parts of marine banks. The Bahama Cays, for example, are located on the Great Bahama Bank, a submarine Tertiary limestone plateau at depths of up to 20 m below sea level. The cays are lined up on the edges of Bahama Bank sections, adjacent to deep marine channels. The banks were exposed during the last glacial sea-level fluctuation (120 m lower), and have been strongly dissected by fluvial erosion. Construction of the cays then occurred together with the Post-Pleistocene sea-level rises. They consist of oolithic limestones, algae-cemented ooide sands, and coral structures. Their development is therefore a result of accumulation as well as of organogenic build-up by corals.

Cat Cays, Bimini Islands, Bahamas

10.25 Atoll

Atolls are circular or elliptic, and occasionally gapped, coral reefs in open oceans. The central area is a lagoon instead of an island. While the width of the reef structure is usually only a few hundred metres, the diameter of the lagoon can vary considerably: between 0·5 and 100 km. The thickness of the coral structures can reach several hundred metres. The structures grow as fringing reefs on a subsiding base formed by eroded and sunken volcanoes. Eustatically rising sea-levels are not the only factors responsible for the development of these forms, as they are not high enough to explain the thickness of the coral structures. The atoll shown here is in the final stage of development as defined in Fig. 29.

Tetiaroa Atoll, Society Islands, French Polynesia (photograph by ZEFA/E. Christian, Vaitape, Bora Bora, Polynesia)

11 Anthropogenic landforms

Artificial forms

Many landforms are the work of humans. Anthropogenic forms were and are being created by the development of settlements, as well as by agricultural and mining activities. Additional forms relate to the construction of transport and water-control structures.

Forms resulting from human settlements have been created since prehistoric times and the remains of earlier settlements often appear as hills or mounds. They are widespread in the Middle East, where they are called tells. These mounds can be of considerable extent and can reach heights of several tens of metres. They were created by the repeated construction of edifices with clay or bricks at the same location, so that many superimposed settlement horizons can be differentiated. Such hills were not only created by the continuing growth of settlements, but were also built as primary structures to control flood damage. This applies to typical features (Wurte, Terpe; Fig. 30, Photo 11.1) on marsh islands along the German North Sea coast, as well as to corresponding anthropogenic hills along the lower reaches and near the mouths of the great rivers of China and India. Prehistoric tombs can also appear as hills, as in Europe with the neolithic and bronze-age grave mound fields (Photo 11.2). Fortification structures from all periods in history are also common. In mountainous terrains, natural forms such as hills or spurs provided convenient locations for the development of fortified settlements. Additional protection was achieved by the construction of moats and walls, which provided the sole means of protection in flat countrysides. An example of a fortification structure is shown in Photo 11.3: a castle built by the Vikings with walls that have survived until today. Old border walls, which also necessitated the moving of large volumes of earth, can still be recognized in the countryside today, if only in parts. One of the most famous examples of border walls is the 548-km-long Roman Limes between the Rhine and the Danube. Another Central European example of a boundary wall is provided by the Danewerk (Photo 11.4).

The use of land for agriculture has created many artificial landforms. A ubiquitous feature in farming areas with abundant rocks in the soil cover are rock accumulations as heaps or walls. Terraced fields are also characteristic anthropogenic forms. They are not restricted to the rice-growing areas of Southeast Asia (Photo 11.5), as former terraced fields can also be observed in higher relief areas in Central Europe, mostly in forest areas where they have not been erased. These relict forms are frequently indicative of medieval agricultural activities. The subdivision of southward-facing slopes into flights of small, walled, terraces is a common feature of the Central European wine-growing regions, but these have frequently been replaced by large-scale terraces (Photo 11.6) in the course of farmland consolidation.

Mining has also formed an extensive range of artificial forms. Excavation with simple tools in prehistoric and medieval times has created pits and waste dumps, which can be recognized in many areas in Europe (Fig. 31). They are also common in those areas of North America and Australia which were subjected to 'gold rushes' during the last century and the early years of this century. Modern, fully-mechanized mining operations form opencast or opencut pits (Photo 11.7) of considerable depth and extent, as for example in the brown coal mining areas of the Lower Rhine. Both open-cast and underground mining generate huge waste dumps (Photo 11.8). The strongest disturbance of natural land surface relief is possibly due to quarrying. Conservative estimates have shown that these activities have affected a total area of at least 1000 km², possibly even several times as much. An estimate of the total mass of rock moved by underground mining operations between 1900 and 1980 is 110 km³. This amount of material is nearly equivalent to the total amount of eroded material deposited in the sea. The subsurface removal of material frequently causes

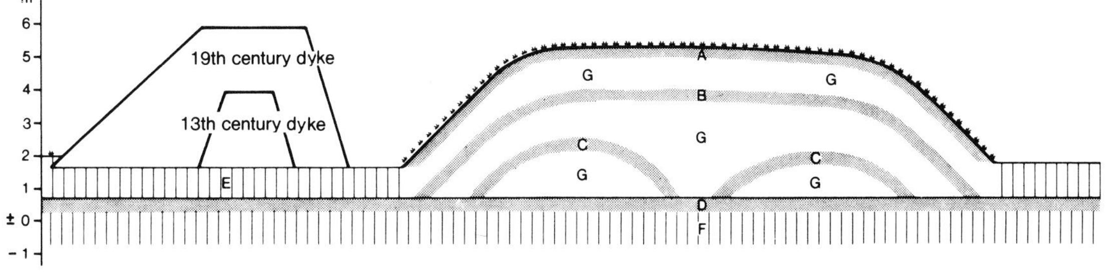

Fig. 30 Schematic section of a settlement mound on the German North Sea coast (from Rathjens 1979)

A: Present humanly-made soil C: Roman humanly-made soil E: Young marsh (tidal flat sediments) G: Mound accumulation material
B: Medieval humanly-made soil D: Natural soil F: Old marsh (tidal flat sediments)

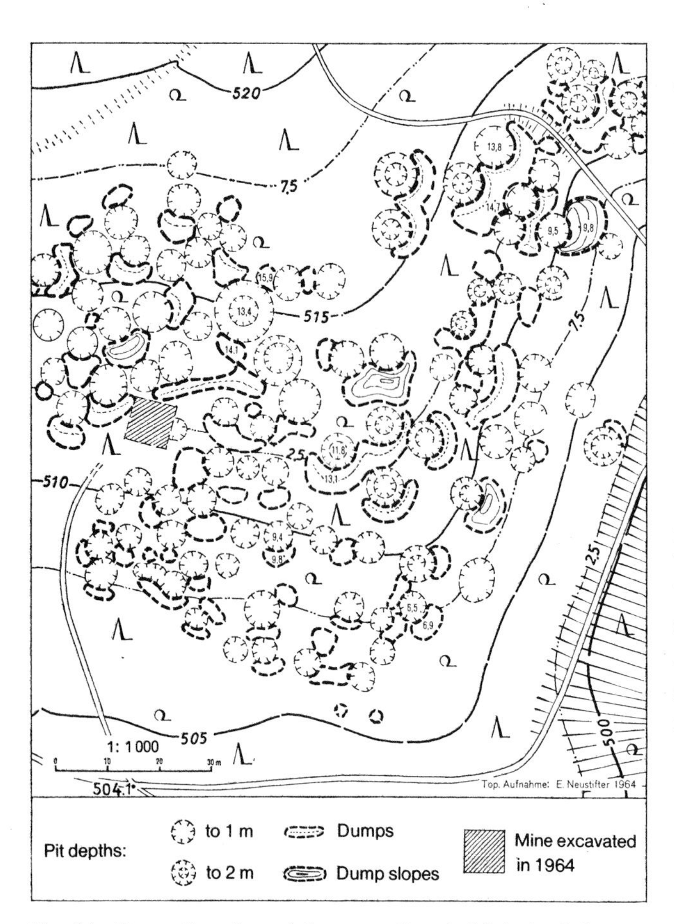

Fig. 31 Excavation pits and dumps northeast of Griesbach, Lower Bavaria (from H. Frei 1966)

Most probably in the Early Middle Ages, between 750 and 1000 AD, iron ore was mined in the Lower Bavarian Tertiary hill country. Many pits up to 10 m deep were excavated in order to obtain the ore. Iron ore was mined from concretion horizons in the Early Pleistocene gravels and the underlying Tertiary Molasse sediments.

damage to surface structures. Underground mining areas are frequently marked by surface hollows caused by subsidence.

Considerable relief modifications are also induced by modern surface transport structures. These include embankments, causeways, and cuttings for both road and rail transport. Medieval roads mostly used higher levels in terrains with strong relief, and often had to surmount steep slopes in order to cross valleys by deep sunken roads. Slopes are often cut by many of these hollows which have been deepened by erosion (Photo 6.9).

Artificial structures have also been created to control flood damage and for irrigation purposes. In mountainous areas, torrents are commonly restructured as a protective measure. Dams or dykes have been constructed along rivers to protect low-lying areas from flood damage. For irrigation purposes, before sprinkler systems were developed dykes, flanking levées along channels, as well as irrigation channels and drainage ditches were necessary. Modern dams (Photo 11.19) are mostly multiple-purpose projects used to regulate water flow for water supply, for the irrigation of agricultural areas, for shipping, for high-water (floodwater) protection, and for electricity generation. Large movements of earth and rock are required for the construction of dams and water channels, as well as for shipping canals (Photo 11.10).

Anthropogenic effects on geomorphological processes

Human influence is not only noticeable in artificial structures. Possibly even more important are the intentional and unintentional ways in which settlement and economic activities trigger or control geomorphological processes. The unintentional initiation of geomorphological processes can most clearly be seen in mass movements and soil erosion. Removal of the natural vegetation frequently provokes slide and glide processes (Photos 2.19, 2.20), even in humid areas of cool-temperate climatic zones where erosion is usually relatively weak (Chapter 5). In mountainous terrain, high potential energy enables anthropogenic influences to destabilize ecosystems. For example in the Alpine areas of Europe, the explosive expansion of ski-tourism during the last three decades has had a deleterious effect on mountain ecosystems. A consequence is the acceleration of geomorphological processes, mainly owing to ski-piste construction. The destruction of the vegetation cover and subsequent loss of slope-stabilizing plant roots, together with soil compaction and subsequent loss of soil porosity and water storage capabilities, leads to increased surface water-flow. This then induces slope erosion, especially rill wash, and changes the flow behaviour of streams by increasing their aggressiveness. Slides, mudslides, and avalanches occur with increasing frequency.

Human interference with natural ecosystems, and the subsequent initiation or amplification of geomorphological processes, is not purely a feature of the last few centuries. Deforestation and subsequent strong soil erosion in Central Europe began during the Neolithic. The accumulation of up to 2-m-thick floodplain clays on the gravel-covered floors of V-shaped valleys during the last ice age is related to deforestation activities during the Late Stone Age and Bronze Age, and Medieval deforestation had similar effects. Destruction of vegetation has led to a global increase in fluvial sediment transport. Modern agricultural methods which require large field areas, and modern road construction greatly increase erosion susceptibility. This is especially critical in the seasonally humid tropics and subtropics, but also applies to arid climatic zones. In all of these areas rainfall occurs in intensive, heavy bursts, which increases its effectiveness in soil erosion. Soil erosion by water occurs where agricultural usage (Photo 11.11) has replaced forest cover (even only light growth), where natural grass cover has been destroyed by overgrazing (Photo 11.12), or where road construction has destabilized slopes (Photo 11.13). The removal of natural vegetation not only leads to sheet and gully erosion by water, but also to wind deflation processes.

Water control measures like river regulation, irrigation systems, land reclamation and coastline protection also affect landforms. River straightening has had an especially deleterious effect by causing strong incision which produces floodplain river terraces that are no longer covered during high-water periods. The River Isar, for example, deepened its bed by 6–8 m within several decades of being straightened, and areas in the city of Munich (English Garden and Viktualien Market) that used to be on the floodplain are now no longer subjected to flooding. Perennial channel irrigation in alluvial plains in warm-arid areas has induced rising ground-water levels and subsequent salinization of soils (Photo 11.14). Technical measures for land reclamation (Photo 11.15), for example by increasing marine sedimentation rates along low-lying coasts in the tidal flat coasts of the North Sea, also exert strong effects on local geomorphological processes.

11.1 Settlement mounds and hills
Wurten are human settlement hills on the German and Dutch North Sea coast. They were constructed as a protective measure against storm tides, before dyke construction commenced at about 1000 AD. Because of eustatic sea-level rise, and subsidence of the underlying marsh ground (i.e. younger tidal flat sediments), their height was repeatedly increased (Fig. 30). Construction material in the hills therefore frequently contains anthropogenic debris. The size of the hills is dependent on the number of structures (farm buildings, churches, cemeteries) in the settlement.

Nordstrandischmoor marsh island, Northern Friesland, Schleswig-Holstein, Germany

11.2 Prehistoric grave mound fields
The construction of grave mounds (grave mound culture) was a typical feature of the middle Bronze Age (1550 to 1200 BC) civilization in Central Europe. Sand or clay is the construction material and extended grave fields are a characteristic feature. The grave field shown here contains approximately 350 circular grave mounds. Their height is about 2 m and their diameter can reach 20 m. Larger structures have, however, been recognized.

Pestrup grave field near Wildeshausen, Oldenburger Geest, Lower Saxony, Germany

11.3 Fortification structure in low-relief area
Substantial earth movements were already undertaken during pre- and early historic times for the erection of fortification structures. The walls of Celtic and Germanic castles in Central Europe, for example Altkönig in the Taunus hills (Celtic) and Altenburg near Fritzlar (Germanic) in Germany, are still well preserved. In the highland areas the metre-high walls are often composed of stone blocks obtained from periglacial slope debris, while in flat areas, fortification walls were constructed of piled-up earth. The castle structure of Trelleborg which was built during the tenth century covers a surface area of 7 hectares and is surrounded by a 6-m-high concentric earth wall with a basal width of 17 m; this was augmented on the inside and outside by palisade fences.

Trelleborg near Slagelse, Island of Seeland, Denmark

11.4 Border wall
The Danewerk is a 17-km-long, east–west trending defensive structure between the North and Baltic Seas, or more precisely, between the Treene river in the west and the Schlei in the east. The Danewerk wall was constructed in the ninth century along the southern border of the Danish empire as a protective measure against the Carolinians and the Wenden. The oldest part is most probably the southern Kograben, a 1-m-high earth wall with a length of 6·5 km and a ditch on its southern side. The wall system was renewed and extended in 1858 by the Danes.

Kograben of the Danewerk, south of Schleswig, Schleswig-Holstein, Germany

11.5 Small terraced fields in traditional rice-growing areas
Rice cultivation requires level fields with elevated margins to retain water, and flights of terraces were therefore constructed on gentle slopes in the traditional rice-growing countries of Asia. On steeper slopes these flights can consist of hundreds of stepped small terraces.

near Pasuruan, Java, Indonesia

11.6 Large terraces in wine-growing areas of Central Europe

Terrace construction on southward-facing slopes is a characteristic feature of wine-growing areas of Central Europe. Many small terraces have been replaced by large-scale terraces in the course of farmland consolidation. These enable more economic cultivation, but are, however, ecologically objectionable. In the Kaiserstuhl and Tuniberg hills on the margin of the Upper Rhine Lowlands, 5–6 large terraces have replaced the former sequence of about 25 to 32 smaller terraces on the slopes. The terraces dip inwards slightly in order to prevent soil erosion, and the elevation differences between the terraces can reach 20 m. All terraces are built in loess soil and the high stability of the soil permits relatively steep slopes between the terrace levels.

near Oberbergen, Kaiserstuhl, Baden-Württemberg, Germany

11.7 Opencast pit

Deep opencast or opencut pits are the characteristic feature of modern surface mining operations in which heavy equipment is used. For example, in the West German brown-coal mining areas, up to 400 m of overburden must be removed before the coal seams with a maximum thickness of 70 m can be reached. The rehabilitation of this mining area was performed by almost complete restructuring of the relief of the Ville area to the west of Cologne, and is now regarded as a model for the successful reclamation of surface mining areas. Opencast mines are not only widespread in coal-mining, but also in ore-mining activities. The photograph shows a section of the complex iron ore opencast pit in the Mesabi Range on the southern margin of the ore-rich Canadian Shield. Waste dumps can be seen in the background.

near Virginia, Minnesota, USA

11.8 Mining waste dump

The deposition of large waste dumps is a typical feature of both surface and underground mining operations. The waste dump shown here consists of material from an underground salt mine and has a circular basal diameter of 800 m, and a height of 130 to 155 m, depending on the elevation of the underlying ground.

Neuhof, southwest of Fulda, Hessen, Germany

11.9 Dam

Dams on rivers represent a considerable modification of natural water flow systems. Sediment accumulation occurs in the dam lake because of restricted flow, although this can be limited by repeatedly releasing water through floodgates in the dam wall. Dam breakage results in a flood wave, with catastrophic effects on settlement and cultivated areas below the dam, as well as geomorphological processes of erosion and accumulation. The dam shown here is 203 m high and was constructed in 1960/63 in a narrow gorge of the River Dez where it cuts through the Iranian Zagros Mountains, just before the river enters the Mesopotamian plains. The dam has a storage capacity of 3340 million m^3, and is utilized for electricity generation. It has also enabled an extension of the irrigated land area in Khuzestan by nearly 100 000 ha.

Dez Valley north of Dizful, Khuzestan Province, Iran

11.10 Canal

Modern water regulation systems redirect water that is stored in dams, or taken from rivers, into areas where it is used for agricultural irrigation or for human or industrial consumption. Transport is nowadays mostly accomplished in subsurface pipeline networks, as well as in open canal systems. In the latter case especially, extensive earth movements are required for the construction of canals and dykes. The canal shown here takes water from a river through a deep, and several metres wide, concrete-walled canal. It accompanies the course of the river but has been built on the slope of the valley, just above the gravel-covered alluvial floodplain.

Piave Valley on the southern margin of the Venetian Alps near Crocetta del Montello, Italy

11.11 Soil erosion caused by deforestation

In areas with seasonally humid subtropical climates (Mediterranean type), the susceptibility to sheet and gully erosion which is already quite strong even with a natural vegetation cover, is considerably intensified by the removal of the natural forest cover. Reforestation on the slopes of the deep erosional gullies is an attempt to inhibit soil erosion.

near Setif, Tell-Atlas, Algeria

11.12 Soil erosion caused by overgrazing
The natural vegetation cover in semi-arid areas of cool-temperate climatic zones is formed by a dense grass mantle (steppe). Overgrazing by cattle creates gaps in the grass cover (foreground left) and subsequent susceptibility to erosion. Even gently-dipping slopes can therefore be covered by a dense network of erosive gullies.

Great Plains near Sundance, Wyoming, USA

11.13 Soil erosion caused by road construction
The construction of roads on steep slopes can destabilize slopes under all climatic conditions, but especially under those with frequent strong, heavy rainfall, i.e. in humid tropical areas. The photograph was taken during road construction.

Northern Range near Blanchisseuse, Trinidad

11.14 Soil salinization caused by irrigation
Salinization can occur in all irrigation areas if drainage is insufficient. In the Punjab area of Pakistan, the introduction of perennial canal irrigation has caused widespread salinization, owing to rising ground water levels. Capillary forces caused by strong evaporation draw saline waters to the surface and the subsequent precipitation of sodium sulphate and chloride at or near the surface has obvious negative effects on the soil structure. The whitish-coloured areas of salinization can be recognized as linear features along the canals, as individual points, or as large patches. Drainage ditch systems similar to those applied in other irrigated areas failed here in the enormous, virtually flat Indus Plain of the Punjab. Some success has been achieved in the battle against salinization by pumping water away from large areas in order to lower the ground-water level.

near Lahore, Punjab, Pakistan

11.15 Land reclamation and coastal protection
Land reclamation in coastal areas with tidal flats is achieved by increasing marine sedimentation rates and thereby creating new land. These measures, together with the construction of dykes around the newly-gained land areas, have turned the German North Sea coast into an artificial coastline. Nowadays, however, coastal protection is the primary goal, instead of land reclamation. In the dyke in the foreground, double rows of poles with pine brushwood in between are erected roughly parallel to the coastline (on the water surface in the middleground). These rows of poles form narrow channels and increase the period during which water movements stagnate during the transition from high tide to low tide, resulting in additional mud deposition. Parallel, 2-m-wide ditches are excavated in the 'fields' bordered by the rows of poles, and the mud that is removed is used to build up the intervening 6-m-wide 'beds'. This system is renewed every few years. These measures promote mud deposition in the ditches and the growth of mud-trapping pioneer plants (*Salicornia*) on the beds. When the height of the beds surpasses the mean high water level, a dense mantle of salt-resistant grass is formed and the new land can be surrounded by dykes. The example shown here covers an area of 1200 ha and was constructed in 1959 (the photograph was taken in Autumn 1971). Unlike other older reclaimed areas that were entirely used for agricultural purposes, only 500 ha of this area will be used for agriculture. Most of it (700 ha) is used as a storage area in order to protect the landward areas from flood waters. The outer dyke (in the background) was constructed on the outer, sandy tidal flats (Chapter 10) in order to include the storage basins in the reclaimed area. The inner margins of the storage basin were subdivided into channels and beds as previously described in order to increase the mud-trapping capacity in the landward areas of the basins. This has led to increased plant growth and a wet biotope has been created which is now a marine bird protection area.

Hauke-Haien-Koog, Northern Friesland, Schleswig-Holstein, Germany

Bibliography

Bird, E. C. F. (1984) *Coasts*. Oxford: Basil Blackwell.
Carson, M. A. and Kirkby, M. J. (1972) *Hillslope form and process*. Cambridge: Cambridge University Press.
Carter, R. W. G. (1988) *Coastal environments*. London: Academic Press.
Chorley, R. J., Schumm, S. A. and Sugden, D. E. (1984) *Geomorphology*. London: Methuen.
Clark, M. J. (ed.) (1988) *Advances in periglacial geomorphology*. Chichester: Wiley.
Cooke, R. U. and Warren, A. (1973) *Geomorphology in deserts*. London: Batsford.
Cooke, R. U. and Doornkamp, J. C. (1990) *Geomorphology in environmental management*. Oxford: Clarendon Press.
Drewry, D. (1986) *Glacial geologic processes*. London: Arnold.
Ford, D. C. and Williams, P. W. (1989) *Karst geomorphology and hydrology*. London: Unwin Hyman.
Francis, P. (1976) *Volcanoes*. Harmondsworth: Penguin Books.
French, H. M. (1976) *The periglacial environment*. London: Longman.
Gerrard, A. J. (1988) *Rocks and landforms*. London: Unwin Hyman.
Gerrard, A. J. (1990) *Mountain environments*. London: Belhaven Press.
Goudie, A. S. (1990) *The human impact on the environment*. Oxford: Basil Blackwell.
Gregory, K. J. and Walling, D. E. (1973) *Drainage basic form and process*. London: Edward Arnold.
Jennings, J. N. (1985) *Karst geomorphology*. Oxford: Basil Blackwell.
Kearey, P. and Vine, F. J. (1990) *Global tectonics*. Oxford: Blackwell Scientific Publications.
Nilsson, T. (1982) *The Pleistocene: Geology and life in the Quaternary*. Dordrecht: D. Reidel.
Ollier, C. D. (1974) *Weathering and landforms*. London: Macmillan.
Ollier, C. D. (1988) *Volcanoes*. Oxford: Basil Blackwell.
Ollier, C. D. (1991) *Ancient landforms*. London: Belhaven Press.
Pethick, J. (1984) *Introduction to coastal geomorphology*. London: Arnold.
Petts, G. E. (1983) *Rivers*. London: Butterworth.
Rice, R. J. (1987) *Fundamentals of geomorphology*. London: Longman.
Richards, K. S. (1982) *Rivers: Form and process in alluvial channels*. London: Methuen.
Selby, M. J. (1985) *Earth's changing surface: An introduction to geomorphology*. Oxford: Clarendon Press.
Sugden, D. E. and John, B. S. (1976) *Glaciers and landscape*. London: Edward Arnold.
Summerfield, M. A. (1991) *Global geomorphology*. Harlow: Longman Scientific and Technical.
Thomas, D. S. G. (1989) *Arid zone geomorphology*. London: Belhaven Press.
Thomas, M. F. (1974) *Tropical geomorphology*. London: Macmillan.
Tinkler, K. J. (1985) *A short history of geomorphology*. London: Croom Helm.
Tricart, J. (1972) *The landforms of the humid tropics, forests and savannas*. London: Longman.
Viles, H. A. (1988) *Biogeomorphology*. Oxford: Basil Blackwell.
Washburn, A. L. (1979) *Geocryology*. London: Edward Arnold.

Sources of Figures

Fig. 1 LESER, H. & W. PANZER (1981): Geomorphologie – Braunschweig (Westermann) Abb. 7, S. 26 und LOUIS, H. & K. FISCHER (1979): Allgemeine Geomorphologie. – 4. Aufl. Berlin, New York (de Gruyter) – Fig. 2, S. 18.
Fig. 2 FRISCH, W. & J. LOESCHKE (1986): Plattentektonik. – Darmstadt (Wissenschaftliche Buchgesellschaft) – Abb. 1–3, S. 6.
Fig. 3 BARSCH, D., O. FRÄNZLE, H. LESER, H. LIEDTKE & G. STÄBLEIN (Hrsg.) (1979) Geomorphologische Karte 1 : 25 000 der Bundesrepublik Deutschland (GMK 25), Blatt 4, Wehr. – Berlin.
Fig. 4 BARSCH, D., O. FRÄNZLE, H. LESER, H. LIEDTKE, G. STÄBLEIN (Hrsg.) (1985): Geomorphologische Karte 1 : 100 000 der Bundesrepublik Deutschland (GMK 100), Blatt 2, Freiburg-Süd. – Berlin.
Fig. 5 LOUIS, H. & K. FISCHER (1979): Allgemeine Geomorphologie, – 4. Aufl., Berlin, New York (de Gruyter). – Fig. 27, S. 140.
Fig. 6 BLUME, H. & G. REMMELE (1989): Schollengleitungen an Stufenhängen des Strombergs (Württembergisches Keuperbergland). – Jber. oberrhein. geol. Ver., N. F., 71: 225–246, Stuttgart – Abb. 3, S. 231.
Fig. 7 LOUIS, H. & K. FISCHER (1979): Allgemeine Geomorphologie. – 4. Aufl., Berlin, New York (de Gruyter) – Fig. 10, S. 56.
Fig. 8 RICHTER, M. & D. (1981): Geologie. – 4. Aufl., Braunschweig (Westermann). – Abb. 26, S. 100, nach: ILLIES, H., ohne Angaben.
Fig. 9 HEIERLI, H. (1984): Die Ostschweizer Alpen und ihr Vorland. Säntismassiv, Churfirsten, Mattstock, Alviergruppe, Appenzeller Molasse. – Sammlung geologischer Führer, 75. – Berlin, Stuttgart (Borntraeger) – Abb. 67, S. 150, – nach: KEMPF, T. (1966): Geologie des westlichen Säntisgebirges. – Beitr. Geol. K. Schweiz, 128, Bern – Fig. 48, S. 66.
Fig. 10 RAST, H. (1987): Vulkane und Vulkanismus. – 3. Aufl., Stuttgart (Enke) – Abb. 11, 12, 14, S. 56, 57, 59.
Fig. 11 BIBUS, E. (1980): Zur Relief-, Boden- und Sedimententwicklung am unteren Mittelrhein. – Frankfurter Geowiss. Arb., Ser. D, 1, Frankfurt – Abb. 47, S. 249.
Fig. 12 DAVIS, W. M. & G. BRAUN (1911): Grundzüge der Physiogeographie. – Leipzig, Berlin (Teubner) – Abb. 81, 82, S. 202, 203.
Fig. 13 SIOLI, H. (1956): Über Natur und Mensch im brasilianischen Amazonasgebiet. – Erdkunde, 10: 89–109, Bonn – Abb. 3, S. 92.
Fig. 14 FISCHER, H. (1986): Erläuterungen zur Geomorphologischen Karte 1 : 100 000 der Bundesrepublik Deutschland (GMK 100), Blatt 6, Koblenz. – Berlin – Abb. 10, S. 47.
Fig. 15 EHLERS, E. (1980): Iran, Grundzüge einer geographischen Landeskunde. – Darmstadt (Wissenschaftliche Buchgesellschaft) – Abb. 5, S. 40; nach: WEISE, O. (1974): Zur Hangentwicklung und Flächenbildung im Trockengebiet des iranischen Hochlandes. – Würzburger Geogr. Arb., 42. Würzburg – Fig. 1, S. 31.
Fig. 16a BLUME, H. (1971): Probleme der Schichtstufen-

landschaft. – Darmstadt (Wissenschaftliche Buchgesellschaft) – Abb. 1, S. 11.

Fig. 16b SCHMIDT, K.-H. (1988): Die Reliefentwicklung der Colorado Plateaus. – Berliner Geogr. Abh., 49. Berlin – Abb. 23, S. 76.

Fig. 17 BLUME, H. & H. K. BARTH (1972): Rampenstufen und Schuttrampen als Abtragungsformen in ariden Schichtstufenlandschaften. – Erdkunde, 26: 108–116. Bonn – Abb. 4, S. 115.

Fig. 18 WILHELMY, H. (1972): Geomorphologie in Stichworten, III, – Kiel (Hirt) – Abb. 3, S. 20.

Fig. 19 GÜLDALI, N. (1970): Karstmorphologische Studien im Gebiet des Poljesystems von Kestel (Westlicher Taurus, Türkei). – Tübinger Geogr. Studien, 40. Tübingen – Abb. 8, neben S. 26.

Fig. 20 BINDER, H., K. H. BLEICH & K. DOBAT (1984): Die Nebelhöhle (Schwäbische Alb). – Abh. Karst- u. Höhlenkunde, Reihe A, 4. 6. Aufl., München – Beilage Höhlenplan.

Fig. 21 LEHMANN, H. (1953): Karst-Entwicklung in den Tropen. – Die Umschau in Wissenschaft und Technik, 1953 (18): 559–562. Frankfurt – Bild 8, S. 561.

Fig. 23 WILHELM, F. (1975): Schnee- und Gletscherkunde. – Berlin, New York (de Gruyter) – Fig. 108, S. 240: nach: BESCHEL, R. (1950): Flechten als Altersmaßstab rezenter Moränen. – Z. f. Gletscherkde. u. Glazialgeologie, 1: 152–161. Innsbruck – Fig. 3, S. 159.

Fig. 24 PETERMÜLLER-STROBL, M. & H. HEUBERGER (1985): Erläuterungen zur Geomorphologischen Karte 1 : 25 000 der Bundesrepublik Deutschland (GMK 25), Blatt 26, Seeshaupt. – Berlin – Abb. 6, S. 34.

Fig. 25 SCHULTZ, J. (1988): Die Ökozonen der Erde. – Stuttgart (Ulmer) – Abb. 31, S. 97; nach: MACKAY, J. R. (1972): The World of underground Ice. – Ann. Ass. Am. Geogr., 62: 1–22, Washington D. C. – Fig. 22, S. 19.

Fig. 27 SCHOU, A. (1968): Basse Bretagnes Kyster. – Geogr. Tidskrift, 67: 200–229. Kopenhagen – Fig. 11, S. 220.

Fig. 28 HEMPEL, L. (1985): Erläuterungen zur Geomorphologischen Karte 1 : 100 000 der Bundesrepublik Deutschland (GMK 100), Blatt 4, Esens/Langen. – Berlin – Abb. 4, S. 39.

Fig. 29 RAST, H. (1987): Vulkane und Vulkanismus. – 3. Aufl., Stuttgart (Enke) – Abb. 22, S. 79.

Fig. 30 RATHJENS, C. (1979): Die Formung der Erdoberfläche unter dem Einfluß des Menschen, Grundzüge der Anthropogenetischen Geomorphologie. – Stuttgart (Teubner) – Abb. 3, S. 27.

Fig. 31 FREI, H. (1966): Der frühe Eisenerzbergbau und seine Geländespuren im nördlichen Alpenvorland. – Münchner Geogr. Hefte, 29. München – Abb. 6, S. 26.

Karte auf dem vorderen und hinteren Vorsatz: nach: HAGEDORN, J. & H. POSER (1974): Räumliche Ordnung der rezenten geomorphologischen Prozesse und Prozeßkombinationen auf der Erde. – Geomorphologische Prozesse und Prozeßkombinationen in der Gegenwart unter verschiedenen Klimabedingungen. – Abh. d. Akad. d. Wiss. Göttingen, Math.-Phys. Kl., III. Folge, 29: 426–439. – Göttingen – farbige Kartenbeilage; und TROLL, C. & KH. PAFFEN (1964): Karte der Jahreszeiten-Klimate der Erde. – Erdkunde, 18: 5–28, Bonn – farbige Kartenbeilage.

Sources for Tables

Table 1 LOUIS, H. & K. FISCHER (1979): Allgemeine Geomorphologie. – 4. Aufl., Berlin, New York (de Gruyter) – S. 224.

Table 2 BARSCH, D. & G. STÄBLEIN (1978): EDV gerechter Symbol-schlüssel für die geomorphologische Detailaufnahme. – Berliner Geogr. Abh., 30: 63–78, Berlin – Tabelle 1, S. 66.

Table 3 Karte auf dem vorderen und hinteren Vorsatz.

Table 4 LESER, H. & W. PANZER (1981): Geomorphologie. – Braunschweig (Westermann) – S. 160.

Index

Bold page numbers indicate photos

Index of localities